Nature's Clocks

The publisher gratefully acknowledges the generous contribution to this book provided by the General Endowment Fund of the University of California Press Foundation.

Nature's Clocks

*How Scientists Measure the
Age of Almost Everything*

Doug Macdougall

⊞

UNIVERSITY OF CALIFORNIA PRESS

Berkeley   Los Angeles   London

University of California Press, one of the most distinguished university presses in the United States, enriches lives around the world by advancing scholarship in the humanities, social sciences, and natural sciences. Its activities are supported by the UC Press Foundation and by philanthropic contributions from individuals and institutions. For more information, visit www.ucpress.edu.

University of California Press
Berkeley and Los Angeles, California

University of California Press, Ltd.
London, England

Library of Congress Cataloging-in-Publication Data

Macdougall, J. D., 1944–
    Nature's clocks : how scientists measure the age of almost
everything / Doug Macdougall.
        p. cm.
    Includes bibliographical references and index.
    ISBN: 978-0-520-24975-2 (cloth : alk. paper)
    1. Geochronometry.   2. Geological time.
3. Radioisotopes in geology.   I. Title.

QE 508. M27 2008
551.7'01—dc22   2007046955

Manufactured in the United States of America

17 16 15 14 13 12 11 10 09 08
10 9 8 7 6 5 4 3 2 1

This book is printed on New Leaf EcoBook 50, a 100% recycled
fiber of which 50% is de-inked post-consumer waste, processed
chlorine-free. EcoBook 50 is acid-free and meets the minimum
requirements of ANSI/ASTM D5634–01 (Permanence of Paper).

*For Gustaf Arrhenius, Harmon Craig, Devendra Lal,
and Henry Schwarcz, great teachers all, who kindled
my interest in isotopes and geochemistry*

CONTENTS

ILLUSTRATIONS

FIGURES

TABLES

ACKNOWLEDGMENTS

My agent, Rick Balkin, first planted the idea for this book; for that, and for his help in seeing it through to completion, I am very grateful. Blake Edgar at the University of California Press made many perceptive suggestions along the way that led to a much improved manuscript. Two readers for the press, Professors R. E. Taylor and Tim Jull, also provided many helpful comments and pointed out various errors and inconsistencies in an earlier version of the manuscript, for which I thank them.

Many people generously provided photographs and illustrations. In particular, I'd like to thank Brian Atwater, Pat Castillo, Paul Hanny, Phil Janney, Sandra Kamo, Jere Lipps, Leonard Miller, Cecil Schneer, and Yuichiro Ueno.

CHAPTER ONE

No Vestige of a Beginning . . .

> If nobody asks me, I know what time is, but if I am asked,
> then I am at a loss what to say.
> Saint Augustine of Hippo, A.D. 354–430

While hiking in the Alps one day in 1991, Helmut Simon and his wife had a disturbing experience: they discovered a body. It was partly encased in the ice of a glacier, and their first thought was that it was an unfortunate climber who had met with an accident, or had been trapped in a storm and frozen to death. Word of the corpse spread quickly, and a few days later several other mountaineers viewed it (see figure 1). It was still half frozen in the ice, but they noticed it was emaciated and leathery, and lacking any climbing equipment. They thought it might be hundreds of years old. This possibility generated considerable excitement, and in short order the entire body was excavated from its icy tomb and whisked away by helicopter to the Institute of Forensic Medicine at the University of Innsbruck, in Austria. Researchers there concluded that the corpse was thousands rather than hundreds of years old. They based their estimate on the artifacts that had been found near the body.

As careful as the Innsbruck researchers were, their age assignment for the ancient Alpine Iceman—later named Oetzi after the mountain

Figure 1. Oetzi, the Alpine Iceman, still partly frozen in ice shortly after his discovery. Two mountaineers, Hans Kammerlander *(left)* and Reinhold Messner *(right)* look on, one of them (Kammerlander) holding a wooden implement probably used by Oetzi for support. Photograph by Paul Hanny / Gamma, Camera Press, London.

range where he was found—was necessarily qualitative. An ax found with the body was in the style of those in use about 4,000 years ago, which suggested a time frame for Oetzi's life. Other implements associated with the remains were consistent with this estimate. But how could researchers be sure? How is it possible to measure the distant past, far beyond the time scales of human memory and written records? The answer, in the case of Oetzi and many other archaeological finds, was through radiocarbon dating, using the naturally occurring radioactive isotope of carbon, carbon-14. (Isotopes and radioactivity will be dealt with in more detail in chapter 2, but, briefly, atoms of most chemical elements exist in more than one form, differing only in weight. These different forms are referred to as isotopes, and some—but by no means all—are radioactive.)

Tiny samples of bone and tissue were taken from Oetzi's corpse and analyzed for their carbon-14 content independently at two laboratories, one in Oxford, England, and the other in Zurich. The results were the same: Oetzi had lived and died between 5,200 and 5,300 years ago (the wear on his teeth suggested that he was in his early forties when he met his end, high in the Alps, but that's another chronology story . . .). Suddenly the Alpine Iceman became an international celebrity, his picture splashed across newspapers and magazines around the world. Speculation about how he had died was rife. Did he simply lie down in exhaustion to rest, never to get up again, or was he set upon by ancient highwaymen intent on robbing him? (The most recent research indicates that the latter is most likely; Oetzi apparently bled to death after being wounded by an arrow.) Fascination about the life of this fellow human being, and his preservation over the millennia entombed in ice, stirred the imagination of nearly everyone who heard his story.

Oetzi also generated a minor (or perhaps, if you care deeply about such things, not so minor) controversy. When he tramped through the Alps 5,000 years ago, there were no formal borders. Tribes may have staked out claims to their local regions, but the boundaries were fluid. In the twentieth century, however, it was important to determine just where Oetzi was found. To whom did he actually belong? Although he was kept initially in Innsbruck, careful surveys of his discovery site showed that it was ninety-two meters (about one hundred yards) from the Austria-Italy border—but on the Italian side. As a result, in 1998 Oetzi was transferred (amicably enough) to a new museum in Bolzano, Italy, where he can now be visited, carefully stored under glacierlike conditions.

Radiocarbon dating is just one of several clever techniques that have been developed to measure the age of things from the distant past. As it happens, this particular method only scratches the surface of the Earth's very long history; to probe more deeply requires other dating techniques. But a plethora of such methods now exists, capable of working out the timing of things that happened thousands or millions or even billions of years ago with a high degree of accuracy. The knowledge that has flowed from applications of these dating methods is nothing short of astounding, and it cuts across an array of disciplines. For biologists and paleontologists, it has informed ideas about evolution. For archaeologists, it has provided time scales for the development of cultures and civilizations. And it has given geologists a comprehensive chronology of our planet's history.

The popular author John McPhee, who has written several books about geology, first coined the phrase "deep time." He was referring to that vast stretch of time long before recorded history and far beyond the past 50,000 years or so that can be dated accurately using radiocarbon. But even though McPhee's phrase is a recent invention, the concept of deep time is not. Without a doubt, it is geology's greatest contribution to human understanding. The idea that geological time stretches almost unimaginably into the past secured its first serious foothold in the eighteenth century, when a few brave souls, on the basis of their close observations of nature, began to question the wisdom of the day about the Earth's age, which was then strongly influenced by a literal reading of the Bible. Today, deep time—and also the "shallow time" of the more recent past—is calibrated by dating methods based on radioactivity. These techniques provide the accepted framework for understanding the history of the universe, the solar system, the Earth, and the evolution of our own species. Without the ability to measure distant time accurately, we would be without a yardstick to assess that history and the many basic natural processes that have shaped it.

For as long as we have written records, there are frequent references to time and its measurement. These have been persistent themes not only for scholars and philosophers, but also for those of a more practical bent. From the earliest times, the sun, moon, and stars were used to mark out days, months, and years—to govern agricultural practice and to formulate rough calendars. Wise men and priests of every culture used an understanding of astronomy to predict the time of a solstice or an eclipse, and sometimes they gained great power and influence from this apparently magical skill. By the time of the Greeks, sophisticated instruments were being produced that accurately traced out solar years, lunar months and the phases of the moon, eclipses, and even the movements of the visible planets.

The technical prowess of the Greek craftsmen who made these instruments is hinted at in written accounts from the time but was only truly realized through an accidental discovery in 1900, when a sponge diver came across an ancient shipwreck near the tiny Greek island of Antikythera. He didn't linger at the site of his discovery because the wreck was disconcertingly littered with bodies. However, later divers found that it was also full of works of art. And among the bronze and marble sculptures from the ship that were eventually assembled at the National Museum in Athens was a nondescript chunk of barnacle-encrusted bronze, partially enclosed in a wooden box. This initially overlooked artifact turned out to be one of the most ingenious and complicated time-telling devices ever constructed; it has even been called the world's first computer. The "Antikythera mechanism," as it is now known, is thought to have been made between 150 and 100 B.C. It comprises more than thirty interconnected and precisely engineered geared wheels that work together as an astronomical calendar. Prior to its discovery, this kind of technology was not thought to have been widely used until about the fourteenth century. It is a marvel of Greek intellectual achievement, and must have been highly valued for the knowledge it imparted about time and the universe. Nothing quite like it appeared for another thousand years or more.

Long before the development of the Antikythera mechanism, however, time, especially as it relates to the history of the world, was an important component of religious beliefs. Early Hindu texts describe multiple cycles of creation and destruction of our world, each lasting 4.32

billion years, which, according to these sources, is just one day in the life of Brahma the Creator. By weird coincidence, that number is quite close to today's most precise measure of the Earth's age. But Brahma's nights are just as long as his days, doubling this number to 8.64 billion years. And each Brahma (there are endless cycles of them) lives for one hundred years, so the age of our world quickly becomes unimaginably large according to this system. Regardless of the exact value, however, it is clear that Hindus are used to thinking about truly deep time—time on a vast scale.

Christians, too, developed a time scale for the Earth, theirs based on the Old Testament of the Bible and exceedingly short compared with that of the Hindus. The best known is the monumental work (over two thousand pages long) by the Irish archbishop James Ussher, published in 1650. Although his conclusion—that the Earth was created on the evening of October 22 in 4004 B.C.—is now often the butt of jokes, Ussher was a serious scholar following in the footsteps of many others who had, over the centuries, tried to piece together a history of mankind based on the Bible. (Ussher's date for the creation of the Earth is usually given as October 23, and it is often said, erroneously, that he stipulated the beginning of the working day, 9 A.M., as the start of it all. But in Ussher's conception of the world's beginning, God wasn't quite so precise. What Ussher actually wrote was, "[The] beginning of time, according to our chronology, fell upon the entrance of the night preceding the twenty-third day of October in the year of the Julian calendar 710." Sometimes "entrance of the night" is taken to mean midnight. So whether Ussher really meant October 22 or October 23 is a matter of interpretation.)

Ussher and his scholarly predecessors believed that the Old Testament provided most of the information they needed to document the entire history of the Earth. This was, at the time, not an unreasonable assumption as there were no other data available to calibrate the world's time scale. Adam was created five days after the Earth was made and was 130 years old when his son, Seth, was born; Seth himself had a son when he was 105; and so on. By adding up lifespans, and making some educated guesses about times when there were gaps, these Old Testament scholars thought they could determine pretty accurately when God created the Earth. Ussher's work was the culmination of this kind of calculation, and it held sway for a very long time; for more than two centuries after his book was published, most Bibles were printed with Ussher's dates displayed prominently in the margins throughout the Old Testament.

But as Ussher worked on his Bible-based time scale for the world, the Enlightenment—the so-called Age of Reason—was dawning in Europe. Although initially closely allied with Christian religious ideals, the Enlightenment inevitably led to the modern scientific approach encompassing observation, experimentation, and hypothesis testing of the physical world, and to a much more secular view of nature. Into this milieu stepped a man whose contributions to our understanding of time are often unappreciated, except perhaps among geologists: James Hutton.

Hutton was born in Edinburgh, Scotland, in 1726, and in his prime he was one of a circle of intellectuals that gave the city its nickname Athens of the North (a much more attractive title than its other nickname, Auld Reekie, which apparently referred either to the foul smell of sewage thrown out of tenement buildings into the narrow streets below, or to the sooty smoke of its coal and wood fires, or maybe even to both). The Edinburgh intellectuals included men such as Adam Smith, James Watt, and David Hume, all of whose work had worldwide impact. Hutton's ideas were equally groundbreaking, although his name is far less widely known today than those of his famous compatriots. He was a global thinker, and he set out to develop a coherent explanation for natural processes on the Earth in the same way that Newton had done before him for the movements of the planets.

For part of his life, Hutton was a gentleman farmer. That experience was crucial for his thinking about the time scales of natural processes, because he observed that the soil on his farm formed—very, very slowly—by erosion of the underlying rocks. He also noted that some of the eroded material was washed into rivers and carried to the sea, where it was deposited as layer after layer of mud and silt and sand. Over long periods of time, through processes that he didn't entirely understand, the buried sedimentary layers hardened into solid rocks. But not all these sedimentary rocks remained on the sea floor. They were found commonly on land, too; in fact, many of the buildings in his native Edinburgh were constructed from blocks of sedimentary sandstone cut out of local quarries. How did they get there? Hutton's solution was that deep burial of the ever-accumulating sediments created heat, often to the point of melting, and when that happened, the whole mass expanded and was thrust up out of the sea to form the hills and mountains of dry land.

Hutton was a creative thinker, but he was also a product of his time. It was the beginning of the industrial revolution, and machines were beginning to take over mechanical tasks. Hutton's view was that the workings of the Earth were not very different from the operations of a machine or an industrial process. (The modern view is similar. What used to be called "geology" is now often referred to as "earth system science," a title meant to emphasize the integrated behavior of Earth processes.) Hutton envisioned an Earth progressing through a natural cycle: erosion of the land, deposition of sedimentary layers in the sea, solidification, heating, and uplift. But history didn't begin or end there; this cycle could be repeated ad infinitum, each step automatically requiring that the next follow. And all the geological processes in these cycles, Hutton understood, took place extremely slowly by human standards. It would require unimaginably long periods of time to erode a landscape, build up thick accumulations of mud and sand, harden them into sedimentary rocks, and finally raise them up out of the sea to where they now stand in the countryside. If such cycles occur over and over again, it would mean that today's landscape is the result of only the most recent cycle. The unimaginably long duration of a single cycle would have to be multiplied many times over to explain the whole of the Earth's history.

Most accounts of Hutton's work assume it was stimulated by direct observation. It is difficult to imagine that his ideas might actually owe more to philosophy than to observation—specifically the philosophy, common in Hutton's time, that nature operates in an unchanging way for the benefit of man and the animal world (the production of fertile soil through processes of erosion being one example). Yet that is what Stephen J. Gould argues in his book *Time's Arrow, Time's Cycle,* noting that Hutton visited several now-famous "Hutton localities" only *after* he had worked out his theory for the Earth. Still, even if he used observations simply to bolster his already-developed theories, it is clear that Hutton was an astute observer. He was among the first to challenge the then-popular idea that granite is produced by precipitation from the sea. Instead, Hutton suggested, it is formed by cooling from a molten state (as we now know to be the case for granite and all other igneous rocks). This idea was based on localities where Hutton observed igneous rocks that demonstrably intruded, liquidlike, into preexisting sedimentary rocks. The reality of such processes neatly fit his theory of burial, heating, and uplift, and it emphasized the very long periods of time necessary for all these processes to operate. One of the places Hutton observed this phenomenon was not far from his home in Edinburgh. Today the site is a mecca for visiting geologists. It can be found easily, just a stone's throw from the Scottish Parliament buildings, on a hillside in the royal estate that is now an enormous park within the city of Edinburgh.

Hutton also recognized that the features geologists refer to as unconformities, which are preserved ancient erosion surfaces, constituted strong evidence that his theory was correct. A sketch drawn by his friend John Clerk (another of the Edinburgh intellectuals, Clerk wrote a classic book on naval warfare and was eventually knighted) shows one of the unconformities Hutton visited near the Scottish town of Jedburgh (see figure 2). The wealth of information contained in this simple image is quite amazing. To the casual observer, it looks like a pretty sketch of a rock outcropping in the countryside, but to Hutton the rock layers told a long and complicated story. It was not as though no other geologists had been to this locality; many had. But Hutton viewed it with fresh eyes, and saw that this one outcrop validated most

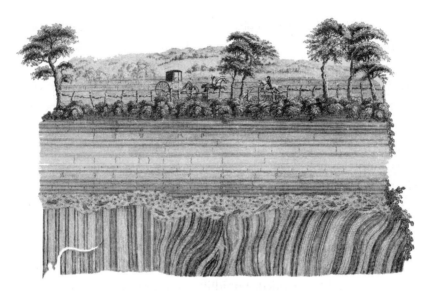

Figure 2. A somewhat idealized sketch of an unconformity observed by Hutton near Jedburgh, Scotland. This sketch, drawn by Hutton's friend John Clerk, appeared in volume 1 of Hutton's *Theory of the Earth, with Proofs and Illustrations,* published in 1795. The sequence of sedimentary layers in this simple drawing illustrates dramatically Hutton's ideas about repeated natural cycles.

of the ideas in his theory. Geology, the evidence in front of him said, is not simply a process of erosion and decay, as some of his compatriots thought. Rather, it involves cycles and includes renewal.

In Clerk's sketch, the lowest band of rock strata stands almost vertical. But because these are sedimentary layers, Hutton knew that originally they had been laid down horizontally in the sea, the accumulated products of erosion of the land, and then buried and hardened into solid rock. Deep burial heated the rocks, and heating led to uplift. Somehow, these once-horizontal rocks had been tilted upright and thrust onto the land. Once out of the protective sea, wind and rain began to take their toll, and erosion produced the slightly undulating surface that can be seen cutting across the upturned strata. This is the actual unconformity, the ancient erosion surface. Note that a layer of loose rubble—unconsolidated erosion products—lies atop the uncon-formity. Hutton's entire natural cycle can be inferred from just this one sequence of rocks. But other sedimentary layers lie above the uncon-formity, these ones horizontal. Their presence requires that the land was once more submerged, sediments again deposited and hardened into rock, and then uplifted (or perhaps the sea retreated), leaving the entire succession once more on dry land. Present-day erosion has formed a layer of soil across the uppermost sedimentary strata. Clerk depicted several human travelers crossing the landscape, presumably blissfully unaware of the great geological story that lay just beneath their horses' hooves.

Hutton's conclusion that the repeated geological cycles required great stretches of time to operate was his most important contribution to sci-ence. Given the prevailing view, even among some scientists, that the Earth was only 6,000 years old, this was a radical idea. There were many critics, and, among other things, Hutton was called an atheist, a slander that in those days was a serious and hurtful charge. Even among those interested in geology and the Earth's history, his ideas were not rapidly accepted; they gained widespread prominence only after they had been popularized by others. Part of the reason was Hutton's writing. While it may have been appreciated by his small circle of fellow intellectuals, it was almost impenetrable to many others, guaranteed to frustrate or put them to sleep. But there is one place where Hutton got it just right. In 1788, in a long paper titled grandly *Theory of the Earth,* he summed up his thoughts about geological time: "The result, therefore, of our present enquiry is, that we find no vestige of a beginning, no prospect of an end." That short phrase—"no vestige of a beginning, no prospect of an end"—has endured; it is as powerful as any that has been written since and is one of the most frequently quoted in all geology.

Hutton's ideas about the immensity of geological time shook up the eighteenth-century world of science and natural philosophy, and the theological world, too. But Hutton did not quantify his results—indeed, at the time he had no way to do so. He didn't know whether the slow geological processes he observed had been going on for a million years, 100 million years, or even longer. His approach was essentially and necessarily qualitative; the task of working out how to measure the time scales of the Earth's operation would have to be carried out by others.

Although it is convenient to treat scientific breakthroughs as singular events, it is rare that they really are so. Hutton is clearly the person who should be credited with establishing the immense sweep of geological time—he was, after all, the first to map out the connections between slow, ongoing processes and the creation of the landscape around us. But there had been earlier rumblings, based on different criteria, that had also suggested a much longer history for the Earth than allowed by the biblical scholars. Even Newton got into the act. He was doing experiments on the rate at which hot objects cool down, and, after determining that a one-inch iron sphere would cool from red heat to room temperature in about an hour, he extrapolated to a sphere the size of the Earth. His calculations indicated that more than 50,000 years would be required. The consensus among Newton's contemporaries was that the Earth had begun its life as a molten globe, and, if this was so, his 50,000-year cooling time would be a rough approximation of its age. Newton never claimed to have determined the Earth's age, but his results were well known among scientists of the time. However, although his estimate was almost a factor of ten greater than Bishop Ussher's 6,000 years, it was still too short to accommodate Hutton's cycles.

More than a century after Newton's experiments, several other researchers used this same approach in explicit attempts to estimate just how old the Earth is. The most famous calculations were done by William Thompson, who was the professor of natural philosophy at Glasgow University for over fifty years, from 1845 until 1899. (Thompson is better known today as Lord Kelvin, a title bestowed on him when he was made a baron in 1892. To avoid confusion, that is how I will refer to him in what follows.) By the time Lord Kelvin did his work on the Earth's age, Hutton's ideas were well entrenched in the geological literature. But Kelvin was a physicist, and he had a physicist's disdain for what he saw as the intuitive and qualitative methods that had been used by Hutton and other geologists. He claimed that Hutton's analysis of the problem was flawed. If the Earth had initially been very hot, or perhaps even molten, he argued, the geological processes in that much hotter past would have been quite different from those we observe today. Hutton had assumed that he could simply extrapolate present-day rates into the very distant past; that, said Kelvin, was wrong.

Why did Lord Kelvin and other physicists think the infant Earth had been very hot? Their main evidence came from observations in deep mines. It was well known that the temperature increases as one descends deeper and deeper into a mine. To a physicist, the existence of such a gradient meant only one thing: our planet is cooling. Heat flowing from a hot interior to the cooler surface produces the observed temperature gradient. This implied a hotter Earth in the past, although just how hot was a matter of conjecture.

Kelvin made some assumptions about the Earth's initial temperature, and about how the process of cooling would proceed, and then calculated how long it would take to reach its present state. He announced his results in 1862: the most probable age for the Earth, he said, was 98 million years. He added a caveat, however. Because of uncertainties in his data and the assumptions he had to make, the actual formation time could lie anywhere between 20 and 400 million years ago.

Lord Kelvin was an influential figure in nineteenth-century Britain, and any results he published were taken very seriously. In addition to his purely scientific work, he was involved in the laying of the first trans-Atlantic cable, and he invented a receiver for the submarine telegraph. Queen Victoria knighted him for his services to science and the country, and the Kelvin temperature scale is named after him. But in spite of his fame, and in spite of the fact that many geologists were chastened by the apparently unimpeachable quantitative approach of this powerful man, there was a lot of unease about his age for the Earth. To some of those who were actively involved in fieldwork and familiar with the everyday processes shaping the landscape, even 98 million years didn't seem to be enough time into which to fit all observable geology.

There was also concern about the very large uncertainty in Lord Kelvin's result—after all, the difference between 20 and 400 million years is huge, a factor of twenty. As a consequence, other scientists, notably a man named Clarence King in the United States, set out to refine the calculations. King accepted Lord Kelvin's assertion that the age of the Earth could be determined by calculating how long it took to cool. However, he also understood that the result of the calculation would only be as good as the data that went into it. It took the invention of the computer to popularize the phrase "garbage in, garbage out," but King understood the principle very well. He knew Kelvin's data on the thermal properties of earth materials—how they held and conducted heat—were not very good, so he set about to improve the situation. He conducted experiments on the melting temperatures of different kinds of rocks, and then extrapolated his data to the high-pressure conditions that prevail in the Earth's interior. With this new information he redid the cooling calculations and concluded that it would have taken just 24 million years for the planet to reach its current state. This was much less than Lord Kelvin's "most probable" age of 98 million years, but it was still within the range he had proposed, albeit near the low end.

Kelvin was pleased because the new result did not contradict his calculations, and he subsequently incorporated King's data into a revision of his own earlier work. By the late 1890s, Kelvin had significantly reduced his allowed range for the Earth's age. It must lie between 20 and 40 million years, he announced, and is most likely closer to 20 than to 40 million. Such was Kelvin's influence that the 20-million-year figure became the accepted wisdom about our planet's age among most scientists. However, this new value caused even more unease among geologists. Not only did they have to fit Hutton's repeated, slow geological cycles into this time span, but now they also had to accommodate the entire course of biological evolution as championed by Charles Darwin.

Lord Kelvin's earlier estimate of 98 million years was already a squeeze; 20 million years did not seem nearly long enough.

Lord Kelvin and Clarence King were by no means the only nineteenth-century scientists to turn their attention to the Earth's age. Nor was the cooling-sphere model the only approach to the problem; many other ingenious ideas were also proposed. Among them was one by George Darwin, the son of Charles and a distinguished scientist in his own right. Darwin assumed that in the beginning the Earth was rotating very rapidly—so rapidly, in fact, that the moon was literally thrown out from the Earth. It was already known in Darwin's day that the Earth's rotation rate is slowly but inexorably decreasing because of tidal friction caused by the moon (and because of this the moon is gradually moving farther away from the Earth). So Darwin calculated how long it would take for the rotation rate to slow to its present value, and came up with an answer of 50 to 60 million years. This, he thought, was a plausible age for the Earth. However, he hedged a bit by saying he couldn't be sure the moon actually formed in this way. If it didn't, it was possible that the Earth was much older.

A completely different but equally imaginative tack was taken by John Joly, an Irish geologist, who made estimates based on the amount of salt in the sea. The source of the salt, Joly knew, is rivers, which continuously carry large amounts of dissolved materials from the continents to the sea. If this process had been going on since the Earth formed, the sea must be getting progressively saltier. Joly reckoned he could calculate the Earth's age simply by dividing the amount of salt in the ocean by the rate at which it is supplied by rivers (he used the sodium content for his calculations; ordinary sea salt is sodium chloride). That sounds straightforward, but Joly, like Clarence King, knew that the result would only be as good as the data used in his calculations. It would obviously be impossible for him to measure the salt content of every river in the world. However, in the best tradition of science, he made reasonable assumptions where he didn't have hard data. His calculations indicated that the Earth is about 90 million years old.

Some geologists tried to determine the Earth's age using an approach that was similar to Joly's, except that they substituted sediments for sodium. But their approach was even more problematic. These scientists had to estimate the total volume of sedimentary rocks that had accumulated over the whole of the Earth's history, and then divide this number by the amount of sediments being formed annually today. Accurately measuring or estimating these quantities was very difficult, and the exercise involved multiple assumptions. Nevertheless, several such calculations were published, and they typically gave ages in the range of 50 to 100 million years. Still, even most of those who had a stake in this work admitted that there were huge uncertainties. And if Hutton was right about recycling, the sediments accumulating today were likely to have been eroded from previously existing sedimentary rock. If this were true, the calculations would substantially underestimate the Earth's age.

In spite of all the caveats, real numbers published in scientific papers are seductive things, and the ages calculated by Clarence King, Lord Kelvin, John Joly, George Darwin, and the geologists tallying up sediment volumes all had their supporters in the scientific community. None of these calculations produced ages greater than about 100 million years, and they ranged down to just 20 million years. These values influenced even geologists who adhered to Hutton's (qualitative) theory of a very ancient Earth. The general consensus was that our planet must be, at most, no more than a few hundred million years old.

Among the early calculations, the estimates made by Clarence King and Lord Kelvin—which gave the youngest values for the Earth's age— seemed to many of their fellow scientists to be the most reliable, because they were based solidly on well-known physical principles. If the Earth had once been hot, and was slowly cooling down, it seemed inescapable that Lord Kelvin's calculations were basically correct. And, indeed, his science was faultless—as far as it went. But neither Kelvin nor anyone else knew then that there are two other natural phenomena that should have been taken into account; their omission made Kelvin's age of the Earth grossly inaccurate. The more important of these phenomena is convection in the Earth's interior, which actively moves hot material toward the surface and cool material to deeper levels. This produces quite a different temperature gradient near the surface than would occur in the rigid Earth that Kelvin assumed for his cooling calculation. The second phenomenon is radioactivity. Small quantities of naturally occurring radioactive isotopes dispersed throughout the Earth's interior produce heat as they decay, and because of this the overall rate of cooling is reduced. In an ironic twist, this same process would, much later, become the basis for our present-day understanding of the Earth's true age.

Radioactivity was discovered very near the end of the nineteenth century. Within less than a decade, several perceptive scientists had realized that it might be a tool for measuring deep time, and a few initial attempts were made to determine the age of rocks that geologists had, up to that time, described only as "very old." The early measurements were rudimentary, but they implied that some of these samples were as old as half a billion years. This was a revolutionary finding—if it were to prove correct, it would mean that the Earth was really many times older than any of the estimates by previous workers had suggested. As you can imagine, there were many skeptics. Supporters of Lord Kelvin simply couldn't comprehend how the great man's calculations could be so badly wrong. Others were so strongly influenced by the entrenched idea that the Earth was no more than about 100 million years old that they simply could not imagine a much older planet. But gradually, as the phenomenon of radioactivity became better understood and more old rocks were dated, most scientists came to accept that the Earth really must be very ancient. There were a few holdouts who for a long time believed that there must be some flaw in the new dating techniques. But, by the middle of the twentieth century, these voices had been drowned out by the success of the approach. As older and older dates were reported, it really did seem that Hutton's "no vestige of a beginning" might be almost literally true.

Radioactivity often gets something of a bad rap; mention it to most people and they immediately think of the devastation at Hiroshima or the nuclear accidents at Three Mile Island or Chernobyl. And it is certainly true that high levels of radioactivity are very dangerous to human health, as was shown dramatically when a Russian ex-spy was mysteriously poisoned in London, England, in 2006. It turned out that the substance responsible for his horrifying and painful death was a radioactive isotope that most people have never heard of, polonium-210. But there is another side of the coin, too. All around us, in the air we breathe, in the water we drink, and in the ground we walk on, there are small amounts of natural radioactivity. In fact, polonium-210 is one of those isotopes, and there are very small amounts of it in your body and mine. In most places on Earth, the quantities of such isotopes are minute enough that their presence poses no danger. But their widespread occurrence is a huge boon for scientists, because it provides a whole array of natural clocks, ticking away in nearly every natural substance.

Dating objects from the distant past using the principles of radioactivity is today referred to as "radiometric dating," and, unlike earlier times, when most of those who did such work were physicists, there is now an entire subfield of the earth sciences devoted to geochronology, the science of measuring past time. Geochronologists may be chemists or geologists or physicists by training, but they have one overarching goal: the accurate measurement of time. Some are mostly interested in improving instrumentation, others in exploring in detail some particular slice of geological time. Together they have managed to find ways to use almost every radioactive isotope that exists in nature to measure the age of things—from the universe itself to archaeological artifacts only a few thousand years old. It has required a great deal of ingenuity and persistence to develop these methods, but the dating tools are now so well honed that they are taken for granted by almost everybody.

That "taking for granted" attitude was one of the primary reasons for writing this book. Most people don't think twice when they hear that archaeologists have found an artifact and dated it to 9,000 years, or that paleontologists have unearthed the fossil of a strange creature that lived 150 million years ago. They don't pause to wonder just how scientists arrive at such amazing conclusions. And when I quizzed friends and acquaintances—and some bright undergraduate students—about radiocarbon dating, it turned out that they had all heard of it, but, beyond that, their understanding was murky. Most of them didn't realize that radiocarbon dating is not useful for dating rocks, or that it is restricted to a very narrow, very recent portion of past time. As for other dating methods, well, for the most part they were completely ignorant. There is nothing inherently wrong with that—especially in this age of information overload, there are many parts of human knowledge that most of us are ignorant about. But it does seem to me that understanding time, especially how time in the distant past is measured and how our ideas about it have evolved and transformed, is crucial to understanding our own place on this planet Earth.

I have been fortunate enough to spend much of my career doing research in isotope geology and geochronology. For me, and, I dare say, for most scientists, there are few things in life more satisfying than the thrill that comes with discovery. Even if it is a very minor discovery in the overall scheme of things, there is nothing quite like realizing you are the first person to know what you have just found out. In this book I have tried to illuminate some such moments in the development of radiometric dating methods, and I hope they provide a sense of the excitement experienced by the scientists who did this work. Even if you are not personally involved, it is hard not to be inspired by the remarkable creativity and inventiveness of those responsible for working out ways to measure the age of almost every conceivable artifact and object from the far reaches of time.

But before I jump into a discussion of just how that is done, and what scientists have discovered using these techniques, I will provide in chapter 2 a brief introduction to radioactivity and how it was discovered, necessary background for understanding radiometric dating. In that chapter, as elsewhere, I have tried to avoid complex or technical discussions that are more suited to a textbook. However, for those who are interested, I have included additional material in appendix C that expands on some of these technical aspects. These short notes are certainly not meant to be comprehensive, but they do introduce aspects of radioactivity that are not covered in the main text and include details of the equations used to calculate ages for some of the dating methods described in the book.

After exploring radioactivity in chapter 2, I deal at some length with radiocarbon dating in chapters 3 and 4—how it came about, and what some of its important applications are. That, I think, is important, because, of all the dating methods that exist, it is the one most commonly in the public eye. It is also the only one that earned its inventor a Nobel Prize. And its development is a good example of how scientists work, and how one discovery leads to another. Furthermore, radiocarbon dating provides a good general introduction to how it is possible to determine the age of things using radioactivity.

Chapter 5 turns to the other end of the time scale and examines the quest to determine the Earth's age accurately using modern dating methods. Doing that was a singular feat, accomplished just over fifty years ago, and, in spite of many refinements in instruments and procedures since then, the result has been little improved upon. Chapters 6 and 7 focus (mostly) on the realm of deep time, exploring how radiometric dating has transformed the originally qualitative and relative geological time scale into an accurate chronology of the Earth's history, and how the progress of biological evolution has been charted through accurate age determinations. Chapter 8 returns again to radiocarbon dating, and examines some of its more interesting recent applications, including such things as working out the timing of earthquakes in the Pacific Northwest of the United States and dating the Shroud of Turin. In the final chapter I highlight some of the important advances in the field of geochronology, and show how these have led its practitioners into some fascinating new fields of research. For reference at the end of the book are a glossary, appendixes containing a current geological time scale and the periodic table of chemical elements, and a listing of books and articles for further reading.

If all these things whet your appetite to learn more about the Earth's history, this book will have accomplished its aim.

CHAPTER TWO

Mysterious Rays

In the cold Warsaw November of 1891, a young Polish woman, just turned twenty-four, packed up her belongings and boarded a train to Paris. She wanted to study science at the Sorbonne, and, although she did not have much money, she was ambitious and very determined. She knew she could stay with her sister, who had moved to Paris earlier, and (this must seem remarkable to any present-day student struggling to finance his or her education) she could attend the great French university for free. Paris transformed her life; four years after arriving there, she married a well-known French scientist, and within twelve years of stepping off the train in Paris as a complete unknown, she was awarded the Nobel Prize in Physics. Today she is a hero of the French Republic. As you may have guessed by now, her name was Marie Curie. That, at least, is how she is known to the world; Paris transformed not only her life, but her name, too. She was born Marya Salomee Sklodowska in Warsaw in 1867.

Marie Curie was one of a small group of scientists whose work during the last years of the nineteenth century and the early years of the twentieth ushered in the discovery of radioactivity and laid the foundations for the field of nuclear science. The others were her husband, Pierre; another French scientist, named Henri Becquerel; the German physicist Wilhelm Roentgen; and, perhaps the most important of them all, the New Zealander Ernest Rutherford. The work of this eclectic and international collection of scientists had an impact on the world that is still felt today.

Marie Curie is credited with coining the term *radioactivity,* a name she chose because the radioactive materials she studied were characterized by strong radiation, although it was a type of radiation quite different from any previously known. She began her work on radioactivity in 1897 as a project for her doctoral degree, inspired by events that had set off a great buzz in the scientific community around the world: the discovery of various kinds of mysterious "rays." The first such discovery had been made by the German physicist Wilhelm Roentgen, who quite unexpectedly observed highly energetic rays emanating from a piece of apparatus that he had constructed to investigate a completely different phenomenon. Roentgen's rays were not actually associated with radioactive materials, but the story of radioactivity usually starts with his work because it set in motion a whole series of investigations that, within a very short time, led to the discovery of radioactivity and revolutionized our understanding of the atom.

Near the end of the nineteenth century, many physicists were experimenting with electricity. Several had investigated whether electricity could flow through a vacuum, or at least a partial vacuum, by discharging an electric current through a sealed glass tube from which most of the air had been pumped away. One such piece of apparatus was known as a "Crookes tube" after the scientist who first designed it, and characteristically the electrical discharges within it produced (in addition to great, crackling, lightning-like sparks) what the researchers called "cathode rays." We now know these as electrons. In addition, the discharges were accompanied by weird and wonderful lighting effects—faint glows within the tube, and fluorescence where the cathode rays hit its glass walls. Fluorescent lights are a modern and much more sophisticated incarnation of these early experimental devices.

Roentgen was an experimentalist, and built most of his own equipment. He also typically repeated the key experiments of other workers when he began a new investigation. It is that habit that found him working away quietly in his laboratory on November 8, 1895. It was Friday night, his laboratory assistants had already gone home, and Roentgen was working with a Crookes tube, discharging electrical currents through it and observing the results. As an aid to detecting the cathode rays, he had coated a sheet of paper with a fluorescing substance; if he held it close to a small "window" that had been cut into the tube and covered with a thin aluminum sheet, the exiting rays would cause the coated paper to fluoresce.

What came next was completely serendipitous and quite startling for Roentgen. As he prepared his experiments, he covered the Crookes tube in black paper and darkened the laboratory so that any fluorescence would be easily visible. The fluorescing screen he had made was sitting on a bench some distance from the apparatus, and out of the corner of his eye he noticed that it glowed whenever he discharged electricity through the tube. But that was impossible! It was sitting halfway across the room, nowhere near the aluminum "window" in the Crookes tube. Cathode rays were weak; they could not penetrate the walls of the tube, or travel very far through the air. Puzzled, he repeated the experiment—and each time he discharged electricity in the tube, the fluorescent screen across the room glowed. In a rare interview, Roentgen was asked what thoughts went through his mind when he first observed this phenomenon. His reply was instructive: he didn't think, he said; he just investigated. He placed the screen at various angles and distances from the glass tube—and still it glowed with each discharge. He could think of only one possible explanation: a powerful, invisible form of energy must somehow be escaping from his apparatus and making the fluorescent screen glow. It had to be something much stronger than the cathode rays. Having no idea what was causing this fluorescence-at-a-distance, he labeled the phenomenon "X-rays."

Roentgen was flummoxed. The characteristics of light, radiation that is visible to the eye, were well known, but energetic radiation that could pass through opaque materials and affect a distant object was unheard of. For the next several weeks, he barely left his laboratory; for a while he ate and slept there so he could instantly act on any inspiration that came into his head as he tried to unravel the properties of the enigmatic rays. He even barred his assistants and his family from entering—his scientific aides first learned about the discovery almost two months later, when Roentgen unveiled it to the world in a "preliminary communication." His most surreal moment must have come as he tested the penetrating power of the X-rays by placing various materials between the glass tube and the fluorescing screen. Most objects produced a vague image of themselves on the screen. But then he picked up a piece of lead and held it out. To his complete astonishment, the image showed not only the shape of the lead, but also a shadowy outline of the bones in his own hand.

We tend to take X-ray images for granted today, so it is difficult to put ourselves in Roentgen's shoes and imagine the impact the experience must have had on him. He was fifty years old, a widely respected scientist near the peak of his career. But this discovery was so bizarre—rays that "saw" through things previously thought to be opaque, like a block of wood or a human hand—that he began to wonder if he was hallucinating. He worked in secret because he had to be sure his observations were real before he publicized the discovery. Uncharacteristically, he didn't even keep laboratory notes during this period. One of his concerns, quite obvious from transcripts of his lectures and some of his later correspondence, was that he was going mad. In a lecture he gave not long after making his discovery, he said, "[I] still believed that I was the victim of deception when I observed the phenomenon of the ray." Later in the same talk he said, "During those trying days I was as if in a state of shock." Shortly after discovering X-rays, he told his wife that he was working on something that would make people think "Roentgen must have gone crazy." But what finally assured him that he wasn't hallucinating were X-ray images he recorded on photographic plates. These were permanent records, not ephemeral visions like the vague outlines he saw on the fluorescent screen. One of the most famous legacies of those weeks of secretive experiments is a radiograph of his wife's hand (see figure 3). Reportedly, she was terrified on seeing the image. This and other early radiographs not only sealed Roentgen's belief in his work, they also must surely have convinced any doubters about the penetrating power of X-rays.

Images of the bones in a human hand made X-rays an instant sensation. Roentgen rays, as they were initially called, became a very hot topic, both in scientific circles and among the public. Much to his chagrin, Roentgen—quite a modest man—became famous. The news of his discovery spread rapidly, and newspapers around the world carried the X-ray pictures he had made. The medical utility of X-rays was quickly recognized, and more whimsical potential uses—"seeing" through clothing, or through locked doors—popped up everywhere in newspaper cartoons. For his work, Roentgen was awarded the very first Nobel Prize in Physics in 1901. Although a man of quite modest means, he donated his entire winnings—a substantial sum—to his university, and he also eschewed patenting his discovery, believing that it should be available for all to use without restriction (under U.S. laws, at least, he would not have been able to patent this natural phenomenon anyway. But many instruments that create and use X-rays have since been patented).

In addition to generating excitement, Roentgen's discovery prompted a flurry of new research. If electric discharges in a glass tube produced X-rays, perhaps there were other unknown types of invisible radiation still to be found. Scientists followed many lines of inquiry, but, because of the apparent connection between X-rays and fluorescence, much of the research focused on substances that were fluorescent or phosphorescent. It was thought that these materials, in addition to emitting visible light, might also be sources of other kinds of radiation. Obvious targets were the abundant, naturally occurring minerals that fluoresce in the dark after exposure to sunlight. These are the same minerals that are

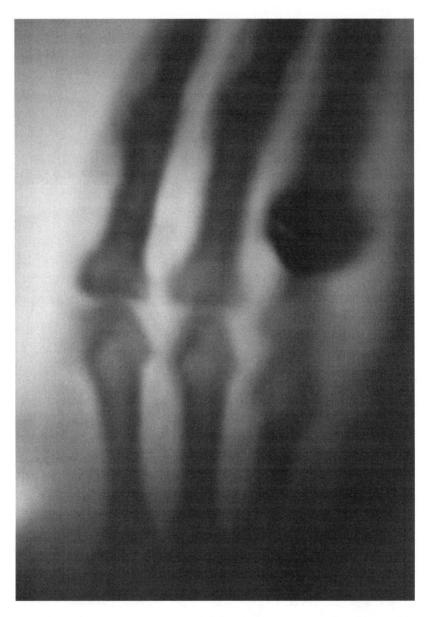

Figure 3. This X-ray image recorded by Roentgen, reputedly of his wife's hand, is almost certainly the first ever made. According to a note with the original photographic plate, it dates to December 22, 1895. Courtesy of the Deutsches Museum, Munich.

often displayed in the geology sections of natural history museums, where they glow in multiple colors when bathed in ultraviolet ("black") light. Many of these fluorescing minerals are compounds of uranium.

Henri Becquerel was a French scientist who had a long-standing interest in fluorescent minerals. He was well connected in scientific circles; his father had been director of the Museum of Natural History in Paris, and Becquerel had worked there as his assistant. When his father died in 1891, Becquerel was appointed professor of physics. As a result, at the time of Roentgen's discovery of X-rays, Becquerel had a well-equipped laboratory in the Natural History Museum, with access to its large collection of minerals and chemical compounds. The combination of his experience working with fluorescing uranium minerals, an interest in the relatively new field of photography, and the resources of the Natural History Museum meant that he was ideally placed to follow up the discovery of X-rays. Becquerel was a talented scientist, but he was also, as the saying goes, in the right place at the right time.

Within months of Roentgen's discovery, Becquerel found that uranium-containing samples affected photographic plates, just as X-rays did. His experimental procedure was simple: he would seal a photographic plate in light-tight black paper and put it on a windowsill. On top of the sealed plate he would place the fluorescent mineral that he wished to investigate. Sunlight would induce fluorescence in the sample, and, he reasoned, if invisible penetrating radiation like Roentgen's X-rays accompanied the fluorescence, it would be detected by the photographic plate. And, indeed, that is what he found. Invariably, after a few days' exposure, a vague image of the sample would appear when he developed the plate. Control experiments with no sample present showed nothing. Becquerel was convinced that visible fluorescence was a necessary condition for the production of the invisible rays.

But crucially, and quite by accident, he soon uncovered evidence to the contrary. The now-famous story has Becquerel preparing a sample in the depths of the Paris winter, but because the skies were so dreary, he didn't immediately put it on a windowsill. Instead, he stored it away in a dark drawer together with its accompanying photographic plate, intending to expose it to sunlight later, when the weather brightened up. For some reason, he never did complete this particular experiment, and the sample remained in the drawer—and therefore never did fluoresce. But he decided to develop the photographic plate anyway. To his great surprise, he found that it showed an image of the sample. It was identical to those produced by the samples exposed to sunlight.

Most of Becquerel's experiments were done with uranium minerals, and he found that they were the only ones that produced images on photographic plates. The penetrating rays he had discovered were clearly different from X-rays because they emanated directly from uranium compounds, not from an electrical discharge tube; but, like the X-rays, their nature was a mystery. They quickly became known as "Becquerel rays," and later "uranium rays." Although he didn't understand the phenomenon, Becquerel had discovered radioactivity. But, for a long time, in spite of his own experimental results to the contrary, he clung to the idea that the rays were somehow connected with fluorescence. Perhaps, he rationalized, they didn't require visible fluorescence, but instead accompanied some sort of unobservable, residual fluorescence. His stubbornness in facing up to the evidence is a good illustration of how even experienced experimentalists can sometimes be swayed by preconceived ideas.

When they learned of Becquerel's work, many other scientists were similarly convinced about the connection between fluorescence and the new rays. They tested every conceivable natural object that emitted light, including things like glowworms and fireflies. A few overly hasty researchers even published papers concluding that these creatures, too, emitted Becquerel rays. But there were problems with their experiments, and—the acid test for scientific ideas—other researchers could not duplicate their results. It soon became clear that it really was only uranium that caused the effects Becquerel had reported.

Like Roentgen, Becquerel gained rapid and widespread fame with his discovery of uranium rays. He wrote a series of scientific papers describing his experiments and results, but then, inexplicably, his interest seemed to wane. In spite of having discovered a completely new phenomenon—an accomplishment for which he, together with Marie and Pierre Curie, was awarded the 1903 Nobel Prize as codiscoverer of radioactivity—he didn't push on with research to understand it better. For a while, it seemed that Becquerel rays were something of a flash in the pan. It was X-rays that commanded most attention, both among scientists and the public. True, a radioactive sample could produce an image on a photographic plate, but compared to X-rays, it took much longer, and the image was much less distinct. It seemed that radioactivity might remain little more than a footnote to the discovery of X-rays.

However, that was soon to change with the work of Marie Curie—closely aided by her husband—and Ernest Rutherford. Unlike Becquerel, these scientists did probe more deeply into the nature of uranium rays. Bit by bit, they lifted the veil on the true nature of radioactivity.

When casting about for a topic to investigate for her doctoral thesis, Marie Curie settled on uranium rays, at least in part because she thought they would be a more fertile area than the already overcrowded field of X-ray research. There is no doubt it was an inspired decision. Her quest brought her great fame (although it also severely affected her health through exposure to high levels of radioactivity). Curie's part in the story of radioactivity has achieved an almost mythical status, particularly in France. In 1995, her remains, together with those of her husband, were removed from their original burial place and interred in the Paris Pantheon in a spectacular public ceremony. She was the first woman to be so honored on her own merits. (One other woman preceded Marie, but it was marriage, not her accomplishments, that got her in. She was the wife of the famous French chemist Marcellin Berthelot.) Even Hollywood wasn't immune to Marie's fame; in 1943, the movie capital released a film documenting her life and achievements. This great public interest has partly to do with her scientific discoveries, but also the fact that her rise to fame had a fairy-tale quality. She was a brilliant young Polish girl who was denied advanced education by the Russian occupiers of her country. She worked as a governess to raise money for her older sister's education in Paris, then later joined her there, living in a garret and earning degrees in both physics and mathematics at the Sorbonne. She continued on to earn a PhD and was awarded the Nobel Prize for her work. Less than a decade later, in 1911, she won a second Nobel Prize, making her the first person to have done so and one of a very select few *ever* to have done so. She became the first woman professor at the Sorbonne, and conceived, organized, and herself drove mobile X-ray units during the First World War to help doctors locate bullets and shrapnel in injured soldiers. For her pioneering accomplishments, she is sometimes referred to as the Joan of Arc of science.

Marie Curie is probably best known as the discoverer of the radioactive element radium; she also discovered a second radioactive element, polonium. These discoveries filled two of the many gaps that still existed in the periodic table of the chemical elements at the end of the nineteenth century. The table, which today hangs on the walls of chemistry classrooms worldwide (and which for reference is included here as appendix B), was devised by the Russian scholar Dimitri Mendeleev in 1869, about thirty years before Marie Curie made her discoveries. Several other chemists independently came up with similar schemes at about the same time, but Mendeleev is the one usually credited because his ideas were so clearly developed. He was a colorful character, usually pictured with a full head of scraggly hair and a long and equally scraggly beard, and he scandalized Russian society by courting and marrying a woman while still married to his first wife. Rumor has it that he threatened suicide if his (second) marriage proposal was rejected. But he was a brilliant chemist.

Mendeleev developed his table while writing a textbook, as a way of classifying the chemical elements (his book became a classic text, both in Russia and elsewhere). He found that if they were arranged in order of increasing atomic weight, their chemical properties changed in a regular way, with periodic repetition of the same properties. By placing them in an array of rows and columns that reflected these changes, Mendeleev showed not only that it was possible to make sense of the chemical properties, but that there were obvious gaps. These were quite numerous; when Mendeleev constructed his first table, only sixty-three elements were known, while modern periodic tables list more than a hundred (some of these are man-made and radioactive). Mendeleev reasoned that there must be undiscovered elements that would fill the gaps, and he predicted their properties from their positions in his table. As new elements were discovered, his predictions proved accurate, and widespread acceptance of his table was assured.

The two radioactive elements discovered by Marie Curie, radium and polonium, fit nicely into Mendeleev's table. Both are unusual because they have short half-lives. (The half-life of radioactive material, as you might guess from the name, is the time required for half a sample to decay away.) The half-lives of radium and polonium are so short, in fact, that you wouldn't expect these elements to exist on Earth at all—they should long since have completely decayed (both elements have several isotopes; their half-lives range from small fractions of a second for some polonium isotopes to 1,600 years for the longest-lived isotope of radium). But unknown to Marie Curie—or anyone else at the time—radium and polonium are part of what chemists now refer to as the "uranium decay series." This is a group of elements that lie between uranium and lead in the periodic table, and all of them are radioactive. They all have short half-lives, and they owe their existence in nature to the continuous radioactive decay of uranium. The uranium decay series has been compared to a row of dominoes—knock one over, and all the others follow. When a uranium atom decays, it is transformed in turn into each of the other radioactive elements in the decay series (including Marie Curie's radium and polonium) until it reaches the nonradioactive end to the series at the element lead.

Marie Curie began her research on Becquerel's mysterious rays by investigating uranium and uranium compounds. Her husband, Pierre, was a gifted experimentalist, and one of the instruments he had developed was a sensitive type of electrometer capable of measuring very small electrical effects. Although he had made it for quite different purposes, Marie suspected that the electrometer would be useful in her own research. It was already well known that there were electrical phenomena associated with the rays, although why that was so was not understood. Marie thought she might be able to determine the amounts of radioactivity in different uranium-containing minerals and compounds by measuring the strength of the electrical charge built up in one of Pierre's electrometers as it was exposed to each sample. This sounds straightforward, but in practice it was difficult—Becquerel himself had tried the same thing and failed. The electrometers were sensitive for their time, but they were primitive by later standards, and it was not easy to measure the very small effects produced by her samples. However, with her characteristic determination, Marie eventually developed the skills and patience needed to complete the crucial experiments—much aided by Pierre's creative instrument tweaking.

Like any good experimentalist entering a new field, once she had perfected her measurement skills, Marie analyzed everything she could get her hands on: minerals from the Museum of Natural History, pure samples of various chemical elements, both liquid (dissolved) and solid versions of some materials, and also uranium ores. She calibrated each sample by comparing its effect on the electrometer with that produced by a known amount of pure uranium.

Her first observation—and this was no great surprise—was that many samples didn't affect the electrometer at all. Only those containing uranium produced a result. This seemed to be an independent confirmation of Becquerel's conclusion—the rays were directly connected with the presence of uranium, in whatever form, and were not associated with other materials. But then came the astonishing part. Marie measured a sample of thorium, and found that it, too, induced a strong response in her electrometer. Thorium occupies a place near uranium in the periodic table, but it is a different element with different chemical properties—and yet her experiments showed that it emitted "uranium" rays! Suddenly she saw Becquerel's rays in a new light; they were *not* just a property of uranium-containing minerals, but a characteristic of at least two different elements—uranium and thorium. Then she measured several samples of the uranium ore pitchblende and found that the electric charge induced in her apparatus was much stronger than she had expected. The ore Marie tested had already been treated to remove uranium for making pottery glazes, but the effect was still several times greater than could be explained solely by the residual uranium content. The only conclusion she could draw was that there must be other elements in the ore, as yet unknown, that emitted the same type of radiation as thorium and uranium. It was at this point that she labeled the new and still very mysterious phenomenon "radioactivity."

Marie Curie's experiments also revealed another fundamental characteristic of radioactivity that distinguished it from normal chemical phenomena: it is purely a property of the element, unaffected by the form of the element or the surrounding conditions. She experimented with different uranium compounds, and found that the strength of the radioactivity always depended only on the amount of uranium in the sample, not on the type of compound. Even if she dissolved a sample in acid and measured it as a liquid, the crucial parameter was still the amount of uranium. Heating or cooling the samples had no effect on the measurements either.

All these observations were communicated to the French Academy of Sciences in April 1898 (the Curies were not members, so their paper had to be presented by Marie's former professor at the Sorbonne). Although it is now recognized as a landmark in the history of radioactivity, the paper initially received little attention. Pierre Curie was a competent scientist, but not among the Paris elite. Marie was just an immigrant, and she hadn't even completed her PhD. Also, she was a woman. How important could their contribution be?

Marie Curie's conclusion that there were unknown radioactive elements in her pitchblende samples was not accepted by many of her fellow scientists. After all, the only evidence she could produce was that the ore was more radioactive than she expected. In every other sample she had examined, the strength of the radioactivity was related to the amount of uranium in the sample, but this correlation didn't hold for pitchblende. That seemed to her critics to be a pretty tenuous argument for the existence of unknown radioactive elements, and Marie realized that the only way to prove she was right would be to separate out these new elements in quantities that could be seen, or weighed, or measured for their chemical characteristics. This turned out to be a monumental task that tried her ingenuity and tenacity. She knew little about chemistry, and nothing at all about the chemical properties of the mysterious elements she set out to separate. It was like finding the needle in a haystack *in the pitch dark*. But, incredibly, and with much trial and error and brute-force chemical processing, in just a year she was able to separate the two new elements, radium and polonium, from the pitchblende. The quantities were minute and the samples impure, but she had enough to be sure about their places in the periodic table. She named polonium after her native country, and radium for the intense radiation it emitted. It was the discovery of these two elements that earned Marie a second Nobel Prize.

Marie was not satisfied, however. She wanted bigger and purer samples of these elements, enough to put in a vial and show the nonbelievers. She decided to concentrate on radium because it was easier to separate from the ore than was polonium, but, even though she now had a better grasp of the necessary chemical procedures, she didn't realize just how difficult the task would be. The work that followed is now part of the Curie legend; it required several years and eventually reached industrial scale as she processed literally tons of pitchblende in an old warehouse. Pierre thought her goal was unattainable, and some scientists thought she was crazy, but she would not be deterred. It took until July 1902, but, in the end, from those tons of pitchblende, she was able to isolate a sample of very pure radium chloride weighing about a tenth of a gram. That is still a tiny amount—less than four one-thousandths of an ounce, and smaller than a drop of rain—but it was enough for her to weigh accurately and thereby determine its atomic weight. It was also

more than enough to convince any remaining skeptics that she had discovered a new element.

As Marie Curie slaved away over her pitchblende, less heroic but no
less significant discoveries about radioactivity were being made in another laboratory on the other side of the Atlantic Ocean—at McGill
University in Montreal, Canada. There, Ernest Rutherford, originally
from New Zealand and like Marie Curie an immigrant, had taken up
the position of professor of physics in 1898. The new laboratory he set up
was devoted to the study of radioactivity.

Rutherford's rise to prominence in science was as meteoric as Marie
Curie's: only three years before his appointment in Montreal, he had left
his native New Zealand as an unknown twenty-three-year-old science
graduate to take up a research scholarship at Cambridge University in
England. But, although unknown, Rutherford arrived in Cambridge
with well-developed skills as an experimentalist, an insatiable curiosity,
and three separate degrees from the University of New Zealand. It
didn't take long for him to impress his Cambridge professors. Again like
Marie Curie, his interest in radioactivity was stimulated by Becquerel's
discovery of "uranium" rays. Throughout his career, Rutherford had a
knack for constructing simple but effective apparatus to help solve
scientific problems. That is exactly what he did at Cambridge, and his
experiments soon showed that Becquerel's rays were actually made up
of at least two quite different kinds of radiation, which Rutherford
called alpha and beta rays. He suspected—correctly, as we now know—
that these "rays" might actually be extremely small particles of matter.
He also detected another type of radiation given off by the uranium:
highly penetrating gamma rays.

Rutherford really wanted to stay at Cambridge University's Cavendish Laboratory, which was a beehive of physics experimentation. But
the prospects of finding a permanent position there in the near term
were dim. A professorship at McGill wasn't quite as prestigious, but as
a second choice it wasn't bad. And he would have the freedom to set up
his own new laboratory. So he accepted the offer, sailed to the New

World, and continued his research on radioactivity with hardly a pause. His tenure at McGill was hugely productive; during a period of just eight years he published some eighty papers, including some of his most innovative work on radioactivity.

Rutherford's research focused on elements that he knew to be radioactive, such as uranium, thorium, and Marie Curie's recently discovered radium. We have already seen that the radium and polonium Curie separated from her pitchblende ore are part of the "uranium decay series" and are present in nature only because they are continually produced by decaying uranium atoms. A closely analogous series exists for the element thorium. A key discovery that Rutherford made, one that led him to a new understanding of the nature of radioactivity, was that one of the radioactive elements in the thorium decay series is a gas.

So what? you might say. Why does it matter whether a particular radioactive element is a gas or a solid? The answer is that gases are very easy to collect in pure form. Marie Curie had to process tons of pitchblende to separate a small amount of radium; all Rutherford had to do was to put some thorium in a sealed container and wait for the gas to accumulate. When he did that, he found that the radioactivity of the gas increased slowly until it eventually reached a steady level, and then stayed the same as long as the thorium was still present. It was in a kind of "steady state," continually being formed by the decay of thorium, but also continually decaying away because it, too, was radioactive. As soon as Rutherford removed the thorium, the radioactivity of the gas began to decrease. However, he then made another, very startling, observation. As the gas decayed away, *another* radioactive element magically appeared in its place. This one was a solid, not a gas. Both the gas and the solid were "new" elements, never before described.

It has to be remembered that Rutherford—and his associate Frederick Soddy, who played an important role in this work—made these observations at a time when atoms were generally thought to be the fundamental, immutable building blocks of matter. And yet their results implied that atoms of one element are actually *transformed* into atoms of another element through radioactive decay. The gas was replaced by a solid. To say the least, this was a stunning discovery. As far as most people were concerned, transformation of one element into another was the domain of alchemy, and alchemy was universally discredited among scientists. Rutherford was acutely aware of this. In his discussions and papers he skirted around the concept of transformation, and took pains to emphasize that his discoveries, radical as they were, didn't really alter the concept of the atom as a primary chemical entity. However, when he was awarded the Nobel Prize in Chemistry in 1908 for his findings, Rutherford—who always considered himself a physicist—quipped that it was the fastest transformation he knew: the award had transformed him from physicist to chemist.

Unlike Marie Curie, who was driven by a need to isolate and characterize the new elements she had discovered, Rutherford was like a kid in a candy shop who wanted to explore every nook and cranny and find every last variety of sweet. Each new observation, it seemed, led to a new experiment. His inquisitiveness was unbounded. Some of his colleagues were convinced that Rutherford had the ability to think on an atomic scale and imagine himself inhabiting the very atoms he was studying. From the starting point of the phenomenon of radioactivity, his work rewrote the textbooks on the atom. When he died, the *New York Times* wrote in a eulogy that "he was universally acknowledged as the leading explorer of the vast infinitely complex universe within the atom, a universe that he was first to penetrate."

What were some of the important findings of Rutherford's experiments? First, as he examined the characteristics of the alpha, beta, and gamma rays he had discovered, he found that they carried tremendous amounts of energy. The alpha and beta rays were actually particles, and they literally shot out of decaying atoms at high speed. The transformation that occurred when an atom underwent radioactive decay released far more energy than did normal chemical reactions. Writing about their experiments in a paper submitted to the journal *Philosophical Magazine* in 1903, Rutherford and Soddy said, "All these considerations [i.e., the results of their experiments] point to the conclusion that the energy latent in the atom must be enormous compared with that rendered free in ordinary chemical change." Rutherford may have been the first person to realize that the atom, properly harnessed, could be a valuable source of energy. He also soon realized, as we will see shortly, that this finding had implications for Lord Kelvin's estimate of the Earth's age.

Secondly, Rutherford's experiments gave him all the information he needed to derive the "law" of radioactive decay, the mathematical description of the way in which the process actually occurs. Fortuitously, the radioactive gas he collected by putting a sample of thorium in a sealed container (we now know the gas is radon) decayed away very rapidly once he had separated it from the thorium. The change in its concentration could therefore be monitored quite easily by measuring the radiation it emitted. Rutherford found that it diminished by about 50 percent each minute, and, after just a few minutes, so little remained that he could no longer detect any radiation. He immediately recognized this as "exponential" decay—that is, the decay occurred at a rate proportional to the amount of radon gas present—a common type of behavior in natural processes. The radon collected in his experiment was characterized by a half-life of one minute, because its radioactivity decreased by one-half every minute. Every other radioactive isotope known also decays exponentially, each characterized by its own half-life. Rutherford had discovered the universal behavior that governs radioactive decay.

It was through his investigations of radioactivity that Rutherford conceived his model of the atom—a model that is, in its broad outlines, not very different from the picture we have today. The set of experiments usually credited with laying the groundwork for this new understanding was designed for quite a different purpose, and, like Roentgen's discovery of X-rays, the results were unexpected. With a little imagination, you can almost put yourself in Rutherford's shoes and feel the thrill he must have experienced as he set up the investigation, and his excitement and amazement at the outcome.

The experiments were designed to examine the properties of the alpha particles Rutherford had discovered in Cambridge. He wanted to investigate their interaction with other kinds of matter, so, in an ingenious arrangement, he put a small sample of radioactive material (which emitted alpha particles when it decayed) at one end of a small tube, and a very thin piece of gold foil at the other. Like bullets fired at a target, the alpha particles sped down the tube with high velocity toward the foil. The particles were very energetic, and he expected they would simply plough through the thin foil. Most of them did. But, unexpectedly, he found that some were deflected away from their straight course. To continue the bullet analogy, it was as though they had ricocheted off something hard and unyielding. Even more startling, a few of the alpha particles were essentially reflected backward. This seemed to Rutherford to be impossible—the foil was too thin and the particles traveling too fast for that to happen. Struggling for an analogy that would illustrate the difficulty, Rutherford said, "It was as though you had fired a fifteen-inch shell at a piece of tissue paper and it had bounced back and hit you." Most of us are not familiar with fifteen-inch shells (I'm certainly not), but you get the picture.

Until Rutherford's experiments, it was generally thought that, whatever the exact nature of atoms, their mass must be spread evenly through matter, including things like the gold foil in his experiments. That is the macroscopic, commonsense view: if you pick up a piece of gold foil (or perhaps aluminum foil would be a better example; most of us don't have rolls of gold foil sitting around the house), it seems to be pretty solid. But Rutherford had to imagine what was happening on an atomic scale. His alpha particles were submicroscopic, and they also carried a positive electric charge. The only explanation that made sense was that the gold atoms of the foil were mostly composed of empty space, with nearly all their mass concentrated in a tiny nucleus that, like the alpha particles, carried a positive electric charge. If that was the case, most of the alpha particles he fired at the foil *should* pass right through, because they wouldn't encounter anything along the way.

A few would have a close encounter with a nucleus, and because both were positively charged, the small alpha particle (atomic weight 4) would be deflected by the much more massive gold nucleus (atomic weight 197). The few alpha particles that met a gold nucleus head-on would be reflected backward. Those were the ones that in Rutherford's analogy "bounced back and hit you."

The concept of the atom we have today owes much to Rutherford's experiments. We know that more than 99.9 percent of the mass of every atom is concentrated in its very small nucleus, and the rest of its volume is nearly empty space, occupied only by a cloud of tiny, almost weightless electrons. It's hard to believe, but the chair I'm sitting on as I write this, the desk that holds my computer—and the book you are reading— are mostly empty space. Once Rutherford understood the structure of atoms, he realized that the phenomenon of radioactive decay must involve the nucleus. The alpha and beta particles, and the gamma rays, must originate there. Even the tiny nucleus must be made up of still smaller components.

Although Rutherford is universally recognized for unraveling the mysteries of the atom, it is less often realized that he almost immediately understood the implications of his work for "radioactive dating." His discovery of the law of radioactive decay, and the fact that each radioactive isotope is characterized by an unchanging half-life, revealed that radioactivity acts as a timekeeper, and that any mineral, rock, or other natural material that contains a radioactive isotope has a built-in clock. Rutherford understood that by measuring how much of the isotope had decayed away or, alternatively, how much of its daughter product had built up in a sample, he could calculate the age of the material.

Never one to miss out on a promising lead, Rutherford got several "old" rocks from some geologist colleagues and set about to determine their ages. He knew from his research that the decay of uranium produces not only the radioactive gas radon, but also another gas: helium. Although he wasn't quite sure how the helium originated (later in his career he would discover that each alpha particle emitted during radioactive decay is actually the nucleus of a helium atom), he knew that it accumulated according to the law of radioactive decay, and, because helium is not radioactive, it doesn't decay away. Old, uranium-containing samples therefore should have high contents of helium, younger samples less. And if he knew the rate at which uranium produces helium as it decays away—a parameter he could measure in the laboratory—he could calculate the age of a sample by measuring its helium content.

Rutherford formally unveiled these ideas in a series of lectures at Yale University in 1905. By then he had made helium measurements on several uranium-rich mineral samples, and he reported that two of these, from quite different localities, gave ages close to 500 million years. It was probable, he said, that these were minimum ages because some of the helium gas might have diffused out of the rocks over their long history.

Obviously, the ages Rutherford reported had to be minimum figures for the Earth's age. And 500 million years was a far bigger number than Lord Kelvin's then still widely accepted 20 million years. But Rutherford had already confronted Kelvin about the age of the Earth while giving a presentation about radioactivity to the Royal Society in London in 1904. Kelvin was in the audience, and Rutherford fretted about how he could tactfully contradict the old man, until he saw him fall asleep partway through the lecture and thought he was home free. However, just as he got to the age of the Earth part, he saw Lord Kelvin's eyes open again. Then Rutherford had a flash of insight. Lord Kelvin, he said, had always claimed that his calculation of the Earth's age would be correct *unless there was an undiscovered source of additional heat in the Earth's interior* (italics mine). Rutherford neatly turned the tables and implied that Kelvin had actually anticipated the discovery of radioactivity. There *was* an additional heat source, he said; it was the radioactive decay of elements like the recently discovered radium, and that meant the Earth would cool much more slowly than Kelvin had calculated. Had he known about radioactivity, Lord Kelvin would certainly have included it as a heat source and would have arrived at a much greater age for the Earth. The elderly Kelvin sat up in the audience and beamed, and Rutherford breathed a sigh of relief. In spite of this incident, however, Lord Kelvin never really accepted the idea of a very ancient Earth. Although he didn't directly dispute Rutherford's results, to the end of his days he believed it was his own work that had given the correct answer.

For most in the scientific community, however, the discovery of radioactivity and the very old ages that Rutherford and a few others measured on uranium-rich minerals quickly consigned Lord Kelvin's ideas about the age of the Earth to history. Not long after Rutherford's lectures at Yale, the minimum age of the Earth was pushed back to over 2 billion years on the basis of helium measurements on a wide range of minerals. Both geologists and physicists joined the race to pin down the age of our planet. There were many twists and turns, and it took many decades before the presently accepted value of 4.5 billion years was reached—a story that is told in more detail in chapter 5.

Today, a wide range of radioactive isotopes, not just those of uranium, are used for age determinations. Nuclear chemists and physicists have catalogued more than sixty naturally occurring radioactive isotopes, and many more that are man-made. They have also found, as Rutherford had inferred, that the nuclei of atoms are themselves composed of even smaller particles. The principal ones, and the only ones we need to be concerned about here, are protons and neutrons. These particles have almost identical (and very small) masses, but the neutron—as its name implies—is electrically neutral, while the proton carries a positive charge. Atoms as a whole are electrically neutral because the negatively charged electrons surrounding the nucleus balance the positive charge on the protons. Radioactivity occurs when the forces that hold these nuclear components together are simply not strong enough and the nucleus becomes unstable. Usually this is because the balance between neutrons and protons is not quite right, or because the nucleus is simply too big, with too many neutrons and protons jostling around inside. When a particle—usually an alpha or beta particle—escapes, the nucleus reorganizes into a more stable configuration. Very large nuclei sometimes just split apart in the process of nuclear fission, which powers the atomic bomb (an unfortunate invention that nevertheless illustrates vividly what Rutherford and Soddy meant when they said that "the energy latent in the atom must be enormous"), but this is not very common.

As Rutherford discovered, one chemical element is transformed into another when alpha or beta particles are ejected. This process changes both the number of neutrons and, crucially, the number of protons in the residual nucleus—crucially, because the number of protons determines the chemical behavior of an element. For example, all atoms that have eight protons in their nuclei are oxygen atoms; all those with twenty-six protons are iron; and so on through the periodic table. The number of neutrons, however, can vary from one atom of an element to another. Oxygen, always with eight protons, may contain eight, nine, or ten neutrons and still be oxygen. Each of these is a different isotope of oxygen, with a different atomic weight (which, in whole numbers, is just the sum of protons plus neutrons in the nucleus). Thus oxygen atoms can have an atomic weight of sixteen, seventeen, or eighteen (eight protons plus eight, nine, or ten neutrons) and are identified as oxygen-16, oxygen-17, or oxygen-18. (For ease of reading in this book, I have followed the convention of writing out isotope names—e.g., carbon-14—rather than using the more usual modern scientific notation, ^{14}C.) All three stable isotopes of oxygen behave the same way chemically, but they can be separated in some processes because of their slightly different weights.

The rate at which a radioactive isotope decays is usually characterized by its half-life, which, as explained earlier, is the time required for one-half a quantity of radioactive atoms to decay. Known half-lives range from incredibly small fractions of a second for very unstable nuclei to billions of years for atoms like thorium. And, because the energy released in radioactive decay is so large relative to that of other natural processes, the rate of decay is unaffected by the environment. It doesn't matter whether radioactive atoms sit on the frigid surface of Mars, are baked in the scorching lava of an active volcano, or are subjected to the immense pressures of the Earth's deep interior; they still decay at the same rate. This constancy is the key to measuring time. Radioactivity is like a clock that never needs adjusting—every second, minute, and hour ticked off precisely, forever. It would be hard to design a more reliable timekeeper.

CHAPTER THREE

Wild Bill's Quest

When the Twenty-ninth International Congress of Americanists was held at the American Museum of Natural History in New York in September 1949, there were three hundred attendees from thirty-five different countries. At the opening session they listened attentively to a talk given by a forty-year-old nuclear chemist from the University of Chicago named Willard F. Libby. It was an unusual venue for a chemist, but Libby's topic was dear to the hearts of the archaeologists in the audience: he spoke about a new method for dating archaeological artifacts that had been developed in his laboratory. "We have reason to believe," he said, "that ages up to 15,000 to 20,000 years can be measured with some accuracy by the present method, and we hope to go to 25,000 years." The next day the *New York Times* reported on Libby's talk. The headline for the article read, "Scientist Stumbles on Method to Fix Age of Earth's Materials." It went on to say, "Experiments with nuclear energy have accidentally uncovered a method . . . of determining the age of Earth's materials." Back in Chicago, Libby's colleagues were not amused. They had been working on the dating idea day and night for most of the previous three years, and they did not appreciate the reporter's implication that it was an "accidental" discovery. One of the scientists had a friend who was a sign painter, and they had him make a

45

fake bronze plaque in metallic paint, which they fixed to the wall of the lab and covered with a window blind. When Libby returned, they asked him to cut the cord and unveil the plaque. It read: "At this point W. F. Libby, age forty, stumbled for three years on carbon-14 dating." Libby took it in good humor. He was single-minded in his quest to perfect radiocarbon dating, but he and the small group of colleagues he had gathered in his laboratory were a closely knit scientific family.

The method Libby developed, radiocarbon dating (also sometimes referred to as carbon-14 dating, or just plain carbon dating), is unique among the techniques that are used to probe past time, because it works only for samples that were once alive. It has been used to date trees, plants, animals, people, and even insects, but, unlike the other methods discussed in this book, it can't be used to date rocks. And it can be used only for objects that are quite young in geological terms, roughly 50,000 years or less. Nevertheless, because it is the one dating method that is widely known, or at least widely recognized, I will discuss it quite extensively. If you've never taken a course in geology—or maybe even if you have—the odds are you've never heard of potassium-argon dating, uranium-lead dating, or rubidium-strontium dating. But radio-carbon dating is another story. Most people have heard about it, and many have a vague idea about how it works. It is frequently in the news because it's often used to work out the age of things that we find deeply interesting—like Oetzi, the Alpine Iceman described in chapter 1. Frequently the subjects of radiocarbon dating studies are directly related to human history. When I tell people I'm a geologist and that I work with isotopes, very often the first question they ask is, "Do you do radiocarbon dating?" I don't; my focus has been on methods that reach farther back into geological history. But the fact that this question recurs is testament to the wide visibility of the radiocarbon method.

In spite of the report in the *New York Times,* Libby's development of radiocarbon dating was anything but accidental. Like many other scientific discoveries, his breakthrough arose through a combination of pure curiosity, the ability to combine knowledge from disparate sources, and a great deal of painstaking experimental work. According to his colleagues, Libby always had an entire file drawer full of ideas to test and projects to develop. Usually he had several on the go at once. When a new PhD student came along, he would pull out one file after another until they settled on a topic. Willard Libby was just "Bill" to friends and colleagues, and inevitably his scattergun approach earned him the nickname "Wild Bill." Especially early in his career, he was famous among his colleagues for doing "quick and dirty" experiments, then publishing the results—and reaching conclusions that were not always correct. But among all of Libby's ideas and accomplishments—and there were many—radiocarbon dating had a special place, and he approached it very seriously and systematically. He knew that if it really worked, it would be groundbreaking, and he pursued its development with singular intensity. His instincts were correct, because it did work, and very well. So well, in fact, that it won him the 1960 Nobel Prize in Chemistry.

In the 1930s and 1940s, new instruments were being developed for detecting and measuring radiation, and research into all aspects of radioactivity was being actively pursued across the United States and elsewhere in the world. Libby was deeply involved in this effort, and he followed new developments in nuclear chemistry closely. He was well aware that nuclear chemists had predicted—purely from theoretical considerations—the existence of many radioactive isotopes that had not yet actually been "discovered." Carbon-14 was one of them. Then, in 1939, Serge Korff of New York University and his colleagues suggested (again from theory) that carbon-14 might be produced naturally in the upper atmosphere. Libby realized almost immediately that if it existed in nature, it might be useful as a dating tool—provided its half-life (then unknown) was long enough. Just a year later, in 1940, the first sample of carbon-14 was produced, artificially, in Berkeley, California, and its half-life was estimated to be many thousands of years. That meant that Libby's dating idea was feasible, and the long and winding road to its realization began.

Carbon has two stable isotopes, carbon-12 and carbon-13. Several radioactive isotopes are also known, but all, with the exception of carbon-14, have very short half-lives and do not occur naturally on Earth. The discovery of carbon-14—its identification, as opposed to the prediction of its existence by theorists—was made by the chemist Martin Kamen, who worked at the Radiation Laboratory, familiarly known as the Rad Lab, on the campus of the University of California at Berkeley. He was studying photosynthesis, and his idea was to "label" the carbon (an essential element in photosynthesis) in his experiments by replacing some of the stable atoms with radioactive ones. By doing this, he thought, he could monitor the radiation emitted by the tracer atoms and follow their progress through all the steps of photosynthesis. It was a good idea, but there was a problem: the only radioactive isotope of carbon known when he began his work was man-made carbon-11, which has a very short half-life, only twenty-one minutes. When he introduced it into his experiments, it decayed away so quickly that it was no longer detectable after an hour or two, rendering it virtually useless as a tracer. Kamen wondered whether there might be other radioactive isotopes of carbon that he could use, with longer half-lives.

Fortunately for Kamen, many scientists at the Rad Lab were in the business of finding new radioactive isotopes as part of their research in nuclear chemistry and physics. Typically they would use an accelerator to fire protons, or neutrons, or alpha particles at very high velocities at various kinds of "targets." Unlike the comparatively low-energy alpha particles in Rutherford's gold foil experiment, which were deflected by the gold atoms, the accelerated particles traveled so fast and carried so much energy that they could penetrate to the nucleus of the target atoms and initiate a nuclear reaction. These collisions transformed the target nuclei into new isotopes; Kamen hoped that under the right conditions he might be able to produce a long-lived isotope of carbon.

Kamen knew about the predictions for carbon-14, and, even though it might turn out, like carbon-11, to have a short half-life, he thought it would be worth the effort to look for it. He got some pure carbon in the form of graphite and bombarded it using the Rad Lab's cyclotron. Sure enough, after the bombardment there was carbon-14 in the target. The amount was very small, but it was usable—and, based on the slow rate at which the new isotope decayed, its half-life was in the range of thousands of years. It would be ideal as a tracer for his photosynthesis experiments.

Although he couldn't have known it at the time, Kamen's demonstration that carbon-14 could be used as a tracer would revolutionize experimental biochemistry. Carbon is *the* key element in biology, and Kamen showed that, by using a radioactive isotope, its movement could be tracked through complex biological processes and biochemical reactions. And, with accelerators, radioactive tracers could be made for other biologically important elements too. Kamen went on to make many other contributions to science, but his radioactive tracer work is by far his most lasting legacy.

But, you may ask, what does man-made carbon-14 have to do with radiocarbon dating? It is here that Libby's ability to bring together observations from several different fields was especially important. On the one hand, through the scientific grapevine he learned of Kamen's carbon-14 discovery, and particularly that the isotope has a long half-life. On the other, he knew about the predictions that cosmic rays might produce carbon-14 naturally in the Earth's atmosphere. If that were really the case, he asked, what would be the fate of these radioactive atoms? Where would they end up? His conclusion was that they would quickly find themselves part of the organic matter of plants and animals, where, as in Kamen's photosynthesis experiments, they would serve as a kind of natural tracer. With a half-life of thousands of years, this natural carbon-14 might be useful as a chronometer for dating any once-living material. It might be especially useful for archaeology.

The physicists who predicted the formation of carbon-14 in the atmosphere were building on research that can be traced back to 1912, when the Austrian scientist Victor Hess first demonstrated that the Earth is constantly being bombarded by radiation from space. In the early years of the twentieth century, not long after Marie Curie used her husband's electrometers to gauge the strength of radiation emanating from uranium and pitchblende samples, other scientists noted that the most sensitive electrometers seemed to detect radiation almost everywhere. There appeared to be a small but pervasive background of radiation even when there were no known radioactive materials in the immediate vicinity. At first it was assumed that this background radiation came from the tiny amounts of natural radioactivity distributed in rocks and soil, but it was soon found that electrometers registered even higher radiation levels when they were taken to greater elevations—to the top of the Eiffel Tower, for instance, or when they were carried up a hill or mountain.

Hess guessed that the radiation might be coming not from the ground, but from *outside* the Earth. So he and his assistants made a series of balloon flights, taking measurements along the way. Their electrometers had been carefully constructed to withstand the low temperatures and pressures of high altitudes, and they did reach high altitudes—more than seventeen thousand feet—and without using oxygen. They found that the intensity of radiation increased dramatically as they ascended; at the highest levels, it was many times the value on the ground. Hess's hunch that the radiation came from space seemed to be confirmed. In a familiar pattern, his discovery was not immediately accepted with open arms. It took a long time for some scientists to concede that Hess was right, but eventually the skeptics were won over. The radiation originating in space came to be known as "cosmic" radiation, or cosmic rays.

To the uninitiated, "cosmic rays" sound mysterious, something from science fiction or *Star Trek*. But, in fact, they are not at all unusual; they are a common type of matter that pervades outer space. Cosmic "rays" are actually particles—they are the nuclei of atoms—and they continuously bombard the Earth from all directions. Most of these particles originate in stars or in supernova explosions in our galaxy; they are nothing more than the material that makes up these objects, thrown out into space. A similar process happens in our own sun, except that the ejected material is usually called "solar wind" rather than cosmic rays. The atoms are propelled into space and accelerated to such high velocities that their surrounding electrons are completely stripped away. The cosmic ray particles that bombard the Earth are therefore bare atomic nuclei, carrying a positive electric charge.

Like particles accelerated in the Berkeley cyclotron, the cosmic ray particles that smash into their "target"—the Earth's atmosphere—induce nuclear reactions. In the process, they create many different isotopes, some of them radioactive, and one of the more abundant products is carbon-14. As it happens, the carbon-14 is not produced directly by the cosmic rays— rather, it is a so-called secondary product—but that really needn't concern us here. The important points are that carbon-14 is produced in significant quantities, and that the cosmic ray bombardment is responsible.

In the air we breathe, oxygen is the crucial element for life; without it, we would suffocate. However, nitrogen is almost four times more abundant—it makes up 78 percent of the atmosphere. So it is not surprising that the cosmic rays bombarding the Earth—and their secondary products—mostly collide with nitrogen atoms. And because nitrogen sits right next to carbon in the periodic table, it is also not too surprising that many of the collisions produce an isotope of carbon. When neutrons—which are one of the secondary products of the cosmic ray bombardment—hit nitrogen atoms, they make carbon-14. Nuclear chemists write out the process as follows:

$$n + {}^{14}N = {}^{14}C + p$$

which simply means that when a cosmic ray–produced neutron (n) hits a nitrogen-14 nucleus (^{14}N), it knocks out a proton (p), and the nitrogen is transformed into carbon-14 (^{14}C).

The researchers who predicted the formation of carbon-14 by this reaction had estimated that cosmic ray bombardment produces approximately two neutrons every second for each square centimeter of the Earth's surface. Nearly every one of these neutrons collides with nitrogen and makes carbon-14. The net result is the production of about 10 billion billion atoms of carbon-14 every second, averaged over the whole Earth; written another way, this is 10,000,000,000,000,000,000 atoms, a lot of radioactive carbon. As a chemist, Libby knew that these atoms wouldn't just float around forever. They would very quickly combine with oxygen, making carbon dioxide, CO_2. And because the gases of the atmosphere mix rapidly, CO_2 in samples of air from anywhere on the planet should contain the same amount of carbon-14.

That, at least, was Libby's conception. And because plants get their carbon by taking up CO_2 from the atmosphere during photosynthesis, and because animals eat plants, every living thing on Earth should be labeled with that same constant proportion of carbon-14. As long as the intensity of the cosmic ray bombardment had not changed over time, that should be true for plants and animals in the past, too.

Libby eventually outlined his thoughts about radiocarbon in a short scientific paper published in the journal *Physical Review* in 1946, using the carbon-14 production estimate from the cosmic ray researchers to predict how much radiocarbon there should be in living matter. But his paper was purely theoretical; he had not made any measurements. To be confident his calculations were right, he would have to analyze living, biological carbon for its radiocarbon content. This posed a problem, because the level of radioactivity he expected would barely be detectable, if at all, with the instruments then available.

That didn't stop Libby, however. By the 1940s, scientists had found ways to "enrich" samples in specific isotopes. Even though all isotopes of an element have the same chemical properties, they have different weights, and can be separated from one another using processes that depend on this property. The enrichment processes concentrate a specific isotope by removing most of the others; this is how "enriched" uranium (which is enriched in the isotope uranium-235 in comparison to natural uranium) is produced for nuclear weapons. Libby knew about a laboratory in Pennsylvania that was processing carbon to produce isotopically unique material for medical purposes. The procedure was time consuming and very expensive, and it had never been used for the then recently discovered carbon-14. But if it could selectively concentrate the isotope carbon-13 for medical use, the same procedure could also enrich

Libby's samples in carbon-14—assuming, of course, that they contained any. By preconcentrating the radioactive isotope in this way, he would have enough for his instruments to detect and measure. As long as he knew the overall carbon content of the sample prior to enrichment, he could determine the radiocarbon content in terms of the amount of carbon-14 per gram of carbon. (This has become the standard way to report radiocarbon measurements.) Furthermore, the Pennsylvania laboratory used methane—a gas made up of carbon and hydrogen—as the starting material for their process. Libby also knew that a sewage plant in Baltimore, Maryland, produced and collected methane—which is formed by decomposition of biological material—as a by-product. Here was an average sample of contemporary, biological carbon in the form of methane, available in large quantities and in a form that could be put through a proven enrichment process prior to measurement.

Less than a year after he published his ideas in the 1946 *Physical Review* paper, Libby obtained two large samples of methane and had them processed by the Pennsylvania laboratory. One was from the Baltimore sewage plant; the other was natural gas. Why choose natural gas? It was a crucial test, because as a fossil fuel, the carbon it contained was ancient. Any carbon-14 that had been present originally would long since have decayed away. Libby hoped to find carbon-14 in the sewage methane, but if his analyses detected any in the natural gas sample—well, it didn't bear thinking about. It would mean that something was drastically wrong.

The person who actually made the measurements on the preen-riched samples was Ernie Anderson, who started working as Libby's laboratory assistant in May 1946, and went on to become his first graduate student at the University of Chicago (see figure 4). Anderson was excited. The "counters" for detecting the radioactivity of carbon-14 were—by today's standards—very crude. But the enrichment the samples had undergone was huge—six hundred liters of methane gas from the Baltimore sewage plant, just a little less than enough to fill a cube measuring a yard on a side, had been processed to produce a sample that

Figure 4. This 1948 photograph shows Willard Libby *(right)* with his assistant Ernie Anderson *(left)* in the University of Chicago laboratory where all the early radiocarbon work took place. Photograph courtesy of the Special Collections Research Center, the University of Chicago Library.

would go into a counter measured in inches. If Libby's calculation was right, that much methane should contain more than enough carbon-14 for Anderson to detect using the existing counters.

Anderson says that the measurement stands out in his memory as the most important he ever made. When he put that first sample into the counter, it was soon obvious that it contained carbon-14. That in itself was cause for celebration. And when he calculated the results after a long period of counting, the amount agreed closely with Libby's prediction. That meant the theory was basically correct: radiocarbon was being produced in the atmosphere and made its way into plants at the same levels. And that in turn brought radiocarbon dating a step closer to reality.

The sewage methane result would go down in the history of radiocarbon research as the very first measurement of naturally occurring carbon-14. (As you might imagine, it was also a prime target for insider jokes because it had been done on sewage.) It led to research that spawned an entirely new field, and had a huge impact on many others. When the second isotopically enriched sample (the natural gas methane) was measured, no carbon-14 was detected. Everyone breathed a sigh of relief. So far, everything was turning out as expected.

Libby's plan to use radiocarbon for dating was, in principle, very simple. Plants—and through them, animals—acquire a characteristic carbon-14 content while alive, determined by the amount in the atmosphere. But once they die, exchange with the atmosphere ceases, and the radiocarbon gradually decays away. Every half-life, the radiocarbon concentration falls by half, following the exponential law for radioactive decay first discovered by Ernest Rutherford. If, as Libby assumed would be the case, atmospheric carbon-14 had remained constant over many thousands of years, then the age of a once-living sample could be read directly from a graph like that shown in figure 5.

Every living organism today (including you and me) contains around 60 *billion* atoms of carbon-14 for every gram of carbon. That sounds like a lot, and it actually is a great many atoms, but, to put it in perspective, there are an almost unimaginable number of atoms in a single gram of

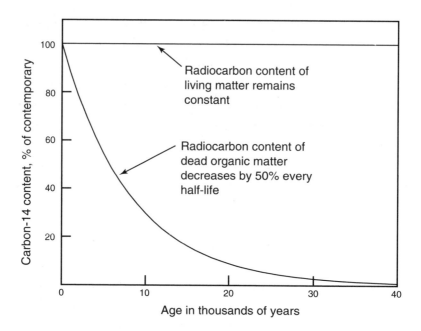

Figure 5. This graph illustrates the principle behind radiocarbon dating. Living animals and plants exchange carbon with the atmosphere and, as long as they are alive, maintain the contemporary carbon-14 value. However, after they die, radioactive decay takes over, and the carbon-14 content decreases exponentially. At 40,000 years, there is very little left.

carbon—picture a number with twenty-two zeros after it. For that reason, only about one in every trillion carbon atoms in your body is carbon-14.

On average, of the 60 billion carbon-14 atoms in every gram of biological carbon, only about 14 decay each minute. Ernie Anderson had been able to measure radiocarbon in the sewage methane at this level only because it had been preenriched, but this process was prohibitively expensive for routine work. When Anderson began his work, the available counters sometimes registered well over a hundred counts per minute even before a sample was inserted. (This was the same background radiation detected by early electrometers, due both to the continuous cosmic ray bombardment of the Earth and to radioactive isotopes in the environment and even in the materials of the counter itself. It is often referred to as "noise.") The radiocarbon in most of the samples for analysis was expected to produce fewer than ten counts per minute. The task seemed impossible. How could you expect to distinguish a signal that was less than 10 percent of the background noise?

Libby, however, was not one to be deterred by such things. He had no doubt the instrument problems could be solved. He gave Anderson two closely related tasks: first, to improve the instruments and then, when he had developed better measurement capabilities, to determine what he called the "contemporary assay," the amount of carbon-14 in contemporary, living biological carbon. The sewage methane was the first such measurement, but, to test whether the value was constant everywhere, Anderson would need to analyze a variety of plants and animals from around the globe. As we will see, this was a crucial step in the development of the method.

In 1946, just as Anderson and Libby were working out how to measure the carbon-14 content of contemporary carbon, a young chemist arrived at the University of Chicago on a one-year postdoctoral fellowship. His name was Jim Arnold, and he had just completed his PhD in chemistry at Princeton, where he had worked on the Manhattan Project (discussed further in chapter 5). Arnold was eager to learn more about radioactivity, and during his year in Chicago he was involved in several different projects with several different advisors. One of them was Bill Libby.

The depth of Libby's conviction that natural carbon-14 could be used for dating was not widely known when Arnold first worked with him. But, when Libby told him about it, Arnold was immediately intrigued. His father was a serious amateur archaeologist with a strong interest in Egypt, and was the American secretary of the British Egypt Exploration Society. Although he was a chemist by training, Arnold had learned about the pharaohs of Egypt from an early age. He could tell you from memory the chronology of their dynasties as it was understood by archaeologists (at the time, this was based on historical records).

When Arnold went home for Christmas that year, he naturally enough told his father about the work he was doing in Chicago. He also mentioned the ideas that were floating around about radiocarbon dating. By then, Ernie Anderson had measured the Baltimore sewage gas, and Arnold probably told his father that to prove the method really worked, Libby would have to analyze samples of known age, and show that radiocarbon dating accurately reproduced the archaeological dates.

The outgrowth of these conversations was unexpected. When Arnold returned to Chicago after the holidays, he found (to his considerable surprise) a box from the Metropolitan Museum in New York containing ten samples from its Egyptian collection. His father, enthusiastic about the fact that his son was involved in a project that might lead to accurate dating of archaeological artifacts, had passed on his enthusiasm to the curator of Egyptian archaeology at the Met. Arnold was somewhat embarrassed because, as far as he knew, at that point radiocarbon dating was still just a pie-in-the-sky idea and very far from reality. But he took the box to Libby anyway, and asked what should be done with the samples. Arnold recalls that Libby didn't enthuse over the samples, or say anything much at all. He also didn't seem particularly surprised; he simply took the box and put it carefully on his shelf. Arnold remembers thinking, for the first time, "My God, he's really serious about this."

Not long after this episode, Arnold's time at Chicago came to an end, and he went off to Harvard and another fellowship. But Libby didn't forget about Arnold, or the samples. There was never any doubt in his mind that radiocarbon dating would be possible, and it was not long before he had secured funding to continue the work. Arnold had been at Harvard for less than a year when he got a call from Libby: Would he consider coming back to Chicago to take charge of the day-to-day work on radiocarbon dating? Arnold accepted without hesitation.

Libby was pursuing the idea of radiocarbon dating with a passion, partly because it seemed to be such a brilliant application of nuclear chemistry, but also because of its potential for age determination in archaeology. That was undoubtedly one of the reasons he asked Arnold to come back to Chicago. The group working on radiocarbon was small, and there was a sense of camaraderie among the researchers. Although there were no rigidly defined boundaries in Libby's laboratory, Arnold assumed much of the responsibility for exploring the dating possibilities, while Ernie Anderson focused on instrument improvement, and eventually on measuring his contemporary assay samples. Unlike some senior scientists, Libby himself remained closely involved, if not always in a hands-on capacity, at least in a supervisory role. He was clearly the driving force and made most of the important decisions, but when it came time to publish results, he nearly always insisted that his young associates should go before him in the author list.

The key advance at this stage of the project was Ernie Anderson's work in reducing the amount of background radiation recorded by the instruments to the point where natural levels of carbon-14 could be measured without prior isotope enrichment. It was a painstaking process; Anderson tried a variety of approaches, such as surrounding the counters with thick lead and iron shields. But his most important breakthrough was to use a method known as "anticoincidence counting." His procedure was to surround the counter containing the sample with a whole series of smaller counters, and then to monitor them all simultaneously. Carbon-14 decays from the sample would register only in the central counter, but background counts from external sources—for example, from cosmic rays—would occur *both* in the central counter and in one of the surrounding ones at the same time. In this way, spurious counts could be detected and eliminated.

Still, it took a great deal of effort to get the rate of the background noise down just to the level of the samples themselves. And, even then, long counting times were necessary to get any useful information from the very small signals. Arnold and Anderson would sit by their instruments and record counts for repeated ten-minute periods, then average the results. They were very aware of statistics—they knew that the actual number of counts would vary considerably from one of those ten-minute periods to the next—but they also knew that the result would be reliable if they averaged measurements from a large-enough number of such periods. However, in the early days at least, they were quite casual in their approach. They knew that the instruments were imperfect, and, if the count rate seemed excessively high or low, they assumed it was a problem with the counter and simply discarded that particular reading. Then they tweaked the instrument and started again. They remember decorating the walls of their laboratory with artwork and slogans. One quotation was from Mark Twain: "There are three kinds of lies: lies, damned lies, and statistics."

With computers, sophisticated electronics, and sleek, commercially produced instruments, modern analytical laboratories can hum along twenty-four hours a day for long periods of time without too much human intervention. In the 1940s, things were very different. The counters that Anderson and Arnold used detected each radioactive decay and produced an audible click. There were no computers to tally these clicks; an operator had to sit next to the instrument with pen and notebook and record them. A great advance in recording technology—and in relieving the boredom of the operators—was the introduction of a strip-chart recorder, a simple device in which a stationary pen registered radioactive decays on a continuously moving strip of paper. At least there would be a permanent record of each count, not so easily subject to human error—and it would mean the counters could run overnight unattended by human operators.

By late 1948, Libby's small group of researchers had made enough progress to attempt a measurement on an archaeological sample. The very first candidate came from the box of Egyptian artifacts Arnold had received from the Metropolitan Museum almost two years earlier. It had been sitting patiently on Libby's shelf, waiting, ever since. The sample they chose was a piece of acacia wood from the tomb of the Egyptian pharaoh Zoser (see figure 6).

Ambrose Lansing, the curator at the Met who had sent the samples to Arnold, had selected them carefully. Each one was well documented archaeologically, and its age was known accurately from historical records.

Figure 6. The very first sample dated using the radiocarbon method came from the tomb of the Egyptian pharaoh Zoser, shown here. Located at Saqqara, near Cairo, the tomb is (for obvious reasons) known as a step pyramid, and is thought to be the first true pyramid built in Egypt. Zoser's burial place is located some 90 feet below ground; the pyramid itself is 204 feet high. Photograph copyright Richard Seaman (www.richard-seaman.com).

However, Lansing had not included a list of dates with the samples because he wanted researchers to do the measurements blind, without prior knowledge of how old they were—a true test of the method. His idea was that Libby would send the results to the Met, and Lansing would then tell Libby whether the age matched the archaeological evidence. But Lansing had overlooked the fact that Jim Arnold had a thorough knowledge of Egyptian archaeology. In fact, Arnold and Libby knew very well the age of that first sample before they made their measurements.

The archaeological evidence dated Zoser's tomb at Saqqara, Egypt, to 4,650 years. Libby and his two young colleagues calculated that for every gram of carbon in 4,650-year-old wood, there should be just over seven carbon-14 decays each minute. But counters are not 100 percent efficient, and, besides, the background levels were still fairly high, so it took a long time to record enough counts from that first sample to be sure about the results. Jim Arnold recalls sitting down after lunch on a Saturday in June 1948 to do the final calculation. He had to subtract the background counts, correct for the efficiency of the counter, and figure out the true count rate for the sample, but when he had done all that, the number staring back at him from the notebook was very close to the value they had predicted. It was what they all had hoped for, but it had by no means been certain it would work out that way. Nature has a way of throwing curve balls, and any number of things could have gone wrong with the experiment along the way. Arnold took a deep breath. He realized at that moment that he was the only person in the world who knew that carbon-14 dating worked. Libby was not at home, and Arnold wandered around for a few sublime hours literally bursting with the news until he could tell someone. "One lives for such moments," he said later.

A repeat measurement on the Zoser sample gave the same result, and when they analyzed wood from another Egyptian tomb from the same time period, they again recorded approximately 7 counts per minute per gram of carbon. The average of all analyses was 7.04 counts per minute, to which they assigned an uncertainty of plus or minus 0.2 counts per minute (more on uncertainties later). These results were even better than they had dared hope for. They had predicted that samples with an age of 4,650 years would give 7.15 counts per minute; in statistical terms, the average of their multiple measurements was indistinguishable from this figure. Everyone in the small Libby group was elated. They were slowly coming to grips with the realization that they had discovered a method that could accurately date an archaeological artifact that was thousands of years old. The results from their measurement of that first archaeological sample were published in the journal *Science* in March 1949, the first formal notice to the world that radiocarbon dating was truly viable.

But a single result, however exciting, would never be enough to establish the technique unequivocally. It was possible, although very unlikely, that the agreement of the radiocarbon dating with the archaeological age of the Zoser sample was a fluke. What they really needed were results from a whole series of samples of known ages. As a first step, Arnold suggested they should measure a sample that was about halfway in age between the Zoser tomb and the present. Being much younger, such a sample should contain substantially more carbon-14 than the wood from Saqqara, and, if their measurements again confirmed predictions, it would be very strong evidence that the Zoser result was not just an unlikely coincidence. Also, Arnold knew, there was an abundance of available material from that part of Egyptian history, the Ptolemaic period. So Libby called Professor John Wilson, head of the Oriental Institute at the University of Chicago and a distinguished Egyptologist, and asked if Wilson could provide a sample of datable material from that age range. Wilson obliged, and sent them a small piece of wood from an artifact in the collection of Chicago's Field Museum.

Arnold began analyzing this second sample with great expectations. But his mood quickly changed because the count rate was far too high, almost twice what he expected. In fact, it was roughly the same as his colleague Ernie Anderson was getting for modern-day carbon. There did not appear to be a problem with the counter; the sample just seemed to contain too much carbon-14 for wood that was several thousand years old. When Arnold calculated the sample's age, he got—within the uncertainties of measurement—zero. Could the sample somehow be contaminated? He discussed the problem with Libby, and they decided to repeat the analysis. But the count rate from the second measurement was similarly high, and the knot in Arnold's stomach tightened. Were they wrong about radiocarbon dating? Was the first result really just a statistically improbable but nevertheless possible coincidence? He analyzed the sample a third time, and again got the same result. By now, more than a month's hard work had gone into these measurements, and, as Arnold remembers, it "spoiled my Christmas." Finally, Libby told him he'd better make an appointment with Wilson to discuss the analyses.

Lab notebook in hand, Arnold sat in Wilson's office and told him that he was really puzzled. Based on his analyses, the sample didn't date from the Ptolemaic period at all; it seemed to be modern. His measurements gave the same result he would expect if he analyzed a piece of wood from a Chicago lumberyard.

"Well, you must be right!" Wilson replied breezily. Arnold, not at all a violent person, says he had "an overwhelming urge to punch this older man right in the face."

The sample that was supposed to be from the Ptolemaic period was a piece of wood from an ancient money case. But, as it turned out, it had not been uncovered in an archaeological dig; it had been purchased in Cairo from a "reputable dealer." It was a fake, and Wilson was not particularly surprised. The sample was the problem, not the dating method. In his defense, Wilson probably didn't realize at the time the seriousness of the Libby dating program, or the agony his faux pas would cause. But he soon learned. Shortly after Arnold reported his conversation with Wilson to Libby, Libby telephoned Wilson and innocently asked him which artifact he would consider to be the most valuable item in the collection of the Oriental Institute. Without hesitation, Wilson replied that it would be the complete throne chair of King Akhenaton, from Egypt's Eighteenth Dynasty, a magnificent artifact displayed in a glass case. "Right," said Libby, "I'm sending Jim Arnold over right away to saw off a leg for radiocarbon dating."

This little anecdote provides an interesting insight into the intensity of effort and personal involvement that went into early radiocarbon measurements. Every sample was shepherded carefully through the measurement process. Today, radiocarbon dating has some of the trappings of a small industry. While many radiocarbon laboratories are dedicated to specific problems, with closely knit groups of scientists working together, there are also laboratories that will date samples for a fee. Archaeologists or geologists can put their fragment of wood or charcoal or bone into an envelope, send it off, and get back a date. In such transactions, much of the excitement and personal stake in the enterprise is lost. The users need dates for their work; those who make the measurements need to ensure that their analyses are technically sound but in many cases have no connection with the samples. And, while it is still true that a lot of thought goes into selecting the appropriate samples for dating, things were quite different during the development of the technique, when every measurement was a potential heartstopper.

The incident of the fake money case made clear to everyone the importance of authenticating each sample, but it didn't slow down the work in Libby's laboratory. By the end of 1949, another paper had appeared in the journal *Science,* authored by Arnold and Libby and including data for six different samples with known ages (they omitted the fake, however; that story wasn't widely known until considerably later). It was this second paper that really brought home to archaeologists and the scientific community in general the huge potential of carbon-14 dating. Part of its impact was undoubtedly due to the way in which the data were presented; the single illustration from that paper, known forever after in the radiocarbon community as the "curve of knowns," has been reproduced in countless other publications, including here as figure 7. It would be hard to find a better example of the old adage that "a picture is worth a thousand words." In their graph, Arnold and Libby plotted the measured carbon-14 content of each of the six samples against its known, historical age. For each point, they included "error bars" that showed the experimental uncertainty of their measurements. On the same graph, they used Ernest Rutherford's law of radioactive decay to draw a curve that showed how carbon-14 would decay away over time (the same curve shown in figure 5), starting at time zero with the "contemporary assay" value that Ernie Anderson had determined for biological material. The key point is that this curve was drawn independently and not fitted to the data points, but it nevertheless passed through five of the six points over an age span of 4,600 years. The single aberrant point that fell slightly off the curve could be explained

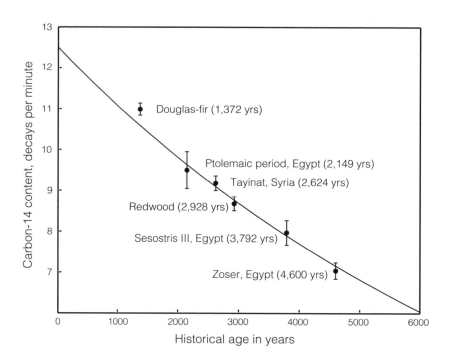

Figure 7. A representation of Arnold and Libby's "curve of knowns." The measured carbon-14 content of each sample is plotted against its age as known from other information (tree ring-counts or archaeological data). The vertical lines, or "error bars," shown with each data point illustrate the experimental uncertainty of the measurement. The descending curve is the decay curve for carbon-14, drawn according to Rutherford's law of radioactive decay, and independently of the data points. Within the measurement uncertainties, all but one of the samples fall on this curve. Based on data in J. R. Arnold and W. F. Libby, *Science* 110 (1949): 678–80.

by statistics. No longer could anyone argue that the agreement was a coincidence. Libby and his colleagues had demonstrated beyond any doubt that radiocarbon dating worked.

As usual, there were still a few who had doubts about the method. Among archaeologists—initially, at least, the most likely to benefit from radiocarbon dating—the doubts were probably partly due to the fact that both Libby and Arnold ·were chemists, not archaeologists. Yes,

Arnold had a good grasp of known chronology in Egypt, but he was not a part of the archaeological community. However, Libby was very astute. He headed off any problems by putting together, early on, an advisory committee comprising several well-known archaeologists and a distinguished geologist. Their charge was to procure and screen samples for radiocarbon analysis—the incident of the fake from the Field Museum was undoubtedly fresh in Libby's mind—and to select a series of important problems for which carbon-14 dating would be especially appropriate. These tasks would have been much more difficult had they been carried out solely by Libby's small group at the University of Chicago, which had no network of colleagues in the geological or archaeological communities. Arnold believes that it was largely because of the advisory committee's work that the first few years of carbon-14 dating were so successful, and that the method gained such rapid and widespread acceptance.

The advisory committee recommended about a dozen areas in which radiocarbon dating could make major contributions, and they began to collect samples for analysis, either from their own collections or by soliciting them from colleagues. From the beginning, it was made clear that the dates would be published. This is not as innocuous as it sounds, because in archaeology and in glacial geology—another area for which radiocarbon dating was to be very important—most of the dates available up to that point were qualitative, sometimes based on little more than instinct or a vague notion about how rapidly some natural process, such as weathering, acted. Anyone with a career staked on a carefully worked-out but untested chronology must have been pretty nervous about handing over key samples for analysis, knowing that they would get back hard numbers that would most likely be taken seriously by their colleagues. They could only bite their fingernails and hope the results would be consistent with their own ideas about chronology.

Fred Johnson, a respected archaeologist who was director of the Peabody Museum at the Phillips Academy in Andover, Massachusetts, headed the advisory committee. Johnson's expertise was in the

Americas; he had worked in the field from northern Canada and Alaska to Mexico. Other committee members were experts in other regions, and among them they encompassed a significant portion of worldwide archaeological interest. The one nonarchaeologist was Richard Foster Flint. Flint was a towering figure, both literally—he was over six and a half feet tall—and in his field of glaciology. Flint's interest was in very recent Earth history, roughly the past 2 million years, during which North America (and other high-latitude regions of the world) repeatedly cycled between long, cold, icy periods and shorter warm intervals like today. The entire period is usually referred to as the Pleistocene Ice Age. Accurate chronology of the glacial cycles was one of the key pieces of information necessary for understanding the causes of the ice age, but before radiocarbon dating, such a chronology was elusive.

With the advisory committee busy soliciting samples—and carefully screening unsolicited samples—and the word getting out about the huge potential of the method, the burden of work in Libby's laboratory was great. Toward the end of 1950, Libby and Arnold decided to publish all the results they had obtained up to that point. This paper, too, appeared in the journal *Science,* in February 1951, and was titled simply "Radiocarbon Dates." It was basically a list describing each sample in detail and giving its age. The list was long; it included about 150 samples. Many had been analyzed twice to increase confidence in the result. "Counting time has been limited to 48 hours," the authors wrote, "in order to accommodate the number of samples necessary to the over-all check of the method, which was the main purpose of this research." They also noted that the dates had been obtained "during the past 18 months." It's easy enough to do the math. Had the instruments been running without a break, twenty-four hours a day, seven days a week, measurement time alone—not counting sample preparation, calculation of the results, and all the other activities that were necessary for each sample—would have amounted to more than a year's effort. Add in the inevitable technical problems with the instruments, and the myriad of other things that can hold up an analysis, and it is a wonder they were able to measure anywhere near that many samples.

Still, it was not all sweat and blood. There were lighter moments, too. Once, when the counters were acting up and things weren't going very smoothly in the laboratory, Jim Arnold wrote a tongue-in-cheek letter to his father, telling him about the problems and suggesting that perhaps the curse of Tutankhamun was responsible. (The tombs of Egyptian pharaohs usually include inscriptions warning intruders of a horrible fate; the curse of Tutankhamun was a media embellishment of this tradition, promulgated when the British financer of the expedition that found Tutankhamun's tomb died just a few months after the discovery. When his dog died shortly thereafter, it only added fuel to the fire.) Arnold senior immediately sent back a small amulet, an "all-seeing Eye of Ra," commonly sold in Egypt to protect its owner from evil (a variant of this symbol, usually referred to as the "Eye of Providence," appears on the back of the U.S. one-dollar bill). After much deliberation about where exactly to place it—inside the shielding, where it could keep close tabs on the counting instruments, or outside, where the scientists would be the ones under its scrutiny—Libby, Arnold, and his colleagues decided that the eye should watch the scientists. Arnold says their results improved immediately.

When Jim Arnold joined the radiocarbon dating project, he told Libby that he didn't want to measure any samples that had strong religious connections. This was purely a pragmatic decision. Cherished religious beliefs are so strong that a "wrong" date—one that contradicted established thinking—could make some people very angry, a complication Arnold wanted to avoid while trying to establish the validity of the method. Radiocarbon dating has since been used widely to date religious artifacts (the story of one particularly famous measurement is told in chapter 8), but, until Arnold left the project, his rule was followed, well, religiously. Almost every other conceivable type of sample was measured, though. Once, when Libby took his group out for lunch at a local restaurant, the conversation turned to shells, probably because there were oysters on the menu. Ordinary shells are made of calcium carbonate and contain a considerable amount of carbon, making them appropriate for radiocarbon dating. However, at the time of the lunch, none had yet been measured. Libby called the proprietor over and asked him where he got the oysters. Chesapeake Bay was the answer. And how did he know? Well, because everyone knew that the best oysters came from Chesapeake Bay, and Morton's Seafood Restaurant served only the best oysters! There was a fair amount of hilarity at this tautology, but before they left, someone in the group scooped up a collection of the shells and took them back to the lab. Chesapeake Bay oyster shells appear as one of the "contemporary assay" samples measured by Ernie Anderson. Another was seal blubber from the Antarctic, which quickly became infamous in the building housing Libby's laboratory because, when they processed it, Anderson says, it smelled like "a skunk magnified."

Even in the earliest phase of research, when Libby's laboratory was the only radiocarbon dating facility in existence, news of the method's potential spread widely. The interest was not confined to scientists. Arnold remembers that there were many curious visitors. For instance, once when Libby was away from Chicago, Arnold was asked to show two prime ministers around the lab. He recalls that one of them, Éamon de Valera of Ireland, was truly interested to learn all the details about the work being done. Within the scientific community, many researchers saw the possibilities of radiocarbon analysis for their own work and wanted to build their own laboratories. A few went it alone, setting up their equipment based on descriptions in the scientific papers published by Libby and his group, or on details they learned about via the scientific grapevine. Libby also actively set out to spread his expertise widely and to recruit people to set up new radiocarbon laboratories. He wasn't interested in keeping others in the dark while he skimmed off the cream, as some might have done in similar circumstances, although it is true that through his advisory group he cornered the market on many interesting samples. On more than one occasion, however, he voiced the sentiment that he "didn't want to be pope" to a whole generation of young archaeologists whose prospects of getting tenure at universities across the country would depend on his measurements of their samples. He wanted them to do the measurements themselves.

Eventually a solution was found for this problem. Jim Arnold set up and ran a kind of minicourse in radiocarbon dating in Libby's laboratory. A whole range of scientists—many of whom would go on to make their own names in this new field—benefited greatly, cycling through the laboratory, taking notes, learning how to operate the equipment, and, as a final test, running a few samples themselves. In very short order, new radiocarbon labs began to spring up across the United States, in Europe, and elsewhere, setting the stage for major advances. Accurate dating was about to rewrite the details of the past 50,000 years of the world's history.

CHAPTER FOUR

Changing Perceptions

In its early years—in the 1950s and 1960s—radiocarbon dating changed people's perceptions of both human and glacial chronology. It didn't actually change the ages of things, of course, but it did change people's understanding of the ages of things, sometimes quite radically. When Libby developed the method, a few other techniques that used radioactivity to probe the chronology of the Earth's past already existed, as will be told later in this book. They, too, were in the early stages of development, and were not yet very sophisticated. But, more important, they were all based on radioactive isotopes with very long half-lives (typically more than a billion years) compared with the 5,730-year half-life of carbon-14. In practical terms, this meant these methods could not access the geologically recent past that became the domain of radiocarbon dating. They simply could not resolve events that had happened during the past 20,000 or 30,000 years. As a result, all the interesting chronology for archaeological and very recent geological events stood on a shaky foundation. Sometimes it involved little more than intuition. Radiocarbon dating quite literally transformed knowledge of this geologically recent swath of time.

Bill Libby's decision to set up an advisory committee to guide his research group in their selection of projects and samples was unusual, but very astute. Unusual because in experimental science, priorities, methods, and sample choices are often jealously guarded from the eyes and ears of potential competitors. Astute because it meant that all the research communities that could benefit from the new method would pay close attention to the results. The committee picked out eleven topics they thought were important and appropriate for radiocarbon dating, and for each they selected a prominent researcher as point man. They circulated details about the research areas and the names of the appointed leaders widely, and anyone who had samples that might contribute valuable age information was urged to contact the committee.

Nine of the research problems they identified were archaeological in nature, and essentially regional in scope, including Peru, the American Southeast, California-Oregon, the Yukon, and Scandinavia, among others. A tenth was geological, involving the timing of advances and retreats of glaciers across Europe, the northern United States, and Canada. The last was "pollen chronology." Pollen grains are produced in vast numbers and spread by the wind, as anyone with a pollen-based allergy knows only too well. But pollen has an upside, too. It is very distinctive—experts can tell what type of plant each pollen grain comes from—and pollen is also remarkably resistant to degradation. In lakes and ponds, pollen grains accumulate along with other sediments (pollen grains accumulate in other places, too, but lakes are especially favorable) and produce a long-term record of the year-to-year regional mix of vegetation. This record in turn is a good indicator of climate, and glacial geologists had been using pollen analysis to track the swings between cold glacial periods and warmer intervals, and to determine whether these changes were local or regional. The problem, however, was to decipher the timing. Radiocarbon dating held out the possibility that accurate ages could finally be deduced for the pollen records, which, until then, had only been useful for determining the relative sequence of events.

The radiocarbon dates produced over the first several years of operation of Libby's University of Chicago laboratory answered many of the questions posed by the advisory committee, and it is not an exaggeration to say that they completely revolutionized the study of both archaeology and glacial geology. Age determination had always been central in these fields, but beyond the time frame of recorded history, many dates were based on informed guesses or, at best, questionable assumptions. In contrast, radiocarbon dates were solidly founded in physics via the law of radioactive decay, and they also could be tested for consistency by multiple analyses of the same sample, or by cross-checking with results obtained by different laboratories. Researchers for whom chronology was crucial became a bit more circumspect in their pronouncements. They realized that both conventional wisdom and off-the-cuff hunches about time scales could now be confirmed by the new dating method—and could also be proved wrong.

One indication of just how powerfully radiocarbon dating affected archaeology came at an international conference in 1990, some forty years after Libby's development of the method. Fred Wendorf, an archaeologist from Southern Methodist University in Texas and a specialist in North Africa, presented a paper in which he said, in part, that "[radiocarbon dating] produced a true revolution in our ideas about the origin and development of almost every known cultural complex [and] profoundly changed our concept of cultural relationships within North Africa, and between North Africa and other areas." He went on to say that radiocarbon dating had rendered obsolete nearly all the chronological relationships *that had been so confidently espoused by the experts* before about 1960 (italics mine). His comments emphasize the importance of reliable and verifiable dates based on radioactivity, as opposed to those obtained in less quantitative ways. His remarks dealt explicitly with archaeology in North Africa, but they are equally valid for other parts of the world.

Carbon-14 dating is now so widespread that there is an entire scientific journal, published three times each year, dedicated exclusively to the results of radiocarbon research. It is called simply *Radiocarbon,* and a typical issue may contain articles about applications in areas as diverse as archaeology, geology, oceanography, and climate change. And that is only the tip of the iceberg. Scientific papers discussing radiocarbon dating appear in many other scholarly journals as well. In the 1950s, however, there was no journal devoted to radiocarbon research. All results from the first few years of work in Libby's laboratory were published in *Science,* the weekly publication of the American Association for the Advancement of Science. *Science* is a prestigious journal with a broad, worldwide readership. Today, researchers vie to have their most important work published there, but the competition is stiff. Many very good papers are rejected, ending up published in other journals that are perceived to have a less exalted status. From that perspective, it is interesting to look again at the early publications by the Chicago laboratory, because today—in spite of their obvious importance—they might not make it into print in *Science.*

The reason is not difficult to grasp. In rapid succession, Libby submitted five separate papers to the journal; they were published between February 1951 and November 1954. The first was coauthored by Jim Arnold and Libby and titled simply "Radiocarbon Dates"; this is the paper described in chapter 3 as reporting ages for about 150 samples. The next four, all with Libby as the sole author, were essentially yearly updates: "Radiocarbon Dates II" through "Radiocarbon Dates V." Each of these papers is little more than a list of dates, together with a detailed description of the samples analyzed. This format was to some extent dictated by the fact that most of the samples had been chosen by the advisory committee, and neither Libby nor his colleagues in Chicago had the expertise to provide interpretations of their results. That would require input from a wide range of specialists. More than 350 different age determinations were dealt with in this way, for samples from geographically far-flung localities. "Just a data dump," today's peer reviewers might say. "No hypotheses to test, no analysis of the significance of the results. This paper should be published in a specialist journal, not in *Science.*"

To be fair, the scientific endeavor has changed drastically since Libby's day. Among other things, there were far fewer options then for publishing data such as the carbon-14 results. And the papers reported results from a new method of potential interest to a broad audience, whom Libby wanted to reach. There is no question that he succeeded in this goal. The published dates were discussed and interpreted widely, both by the specialists who had submitted the samples in the first place and by others in their fields.

The range of materials analyzed for the five papers in *Science* is astonishing. Most common, as you might imagine, are things like charcoal and wood, but also listed are dates for everything from corncobs to human hair, deer antlers, beeswax, and giant sloth dung. Anything that contained carbon and was once alive was fair game. In the second of the series of papers in *Science*—this was after Jim Arnold had left the project—Libby reported a date for the Dead Sea Scrolls. Arnold, as you may recall, had not wanted to work on samples with specific religious significance. The sample Libby analyzed was actually a piece of the linen wrapping of the scrolls, and his result confirmed that the material dated from 2,000 years ago. Another of the papers reports a date for frozen grasshoppers. It's clear that Libby was having fun, enjoying the fruits of having developed a new dating method. The grasshoppers were from a glacier in Yellowstone National Park, and their radiocarbon content was only slightly lower than the contemporary value. Libby calculated their age as 45 years, but, within the fairly large margin of uncertainty in the measurement, they could be anywhere from zero to about 200 years old. Most probably, they were frozen into the glacier during the 1870s or 1880s, when hordes of grasshoppers plagued the western United States.

Some of the dates in the *Science* papers provided only minimum ages, the results given in terms such as "older than 17,000 years" or "older than 25,000 years." In these cases, the counting rates were barely distinguishable from the background rate observed with no sample in the counter, indicating that nearly all the carbon-14 initially present in the sample had decayed. To an outsider, such results might seem of little value. But to anyone seeking to establish an absolute chronology for an archaeological site or a glacial deposit, even a minimum age is a valuable piece of information.

A word about "errors" and "uncertainties" may be useful here. This is a somewhat technical topic, but an important one to understand. Dates reported in the scientific literature are typically given in the form "5,000 years, plus or minus 300 years." This simply means that, to a high degree of probability (which is usually specified precisely when a date is reported), the true age of the sample lies between 4,700 and 5,300 years. It is as likely to be 4,795 years, or 5,123 years, or anything else in that range, as to be exactly 5,000 years. But it is much less likely to fall outside the range. And an age farther away from that range is even less likely than one close to it. The "plus or minus" part takes into account the uncertainty in the data used to calculate the age—for example, uncertainty in the measured count rate in a carbon-14 dating experiment. The usual analogy is coin tossing. We all know that, if you have the patience to toss a coin a million times, the split between heads and tails will be very close to fifty-fifty. The uncertainty of the result in that experiment is small because you have carried out a large number of trials. But, if you toss the coin only three or four times, you might get all heads, or all tails, or some other lopsided result. In this case, the result has a large uncertainty, because the next time you perform the same experiment of three or four tosses, you will be likely to get quite a different result. The more times you perform this experiment, the closer the combined result will be to the correct fifty-fifty split. So, too, with counting radioactive decays from carbon-14 (or any other radioactive isotope). The smaller the number of counts (e.g., for old samples containing little radiocarbon), the higher the uncertainty in the result. Uncertainties, however, are not just related to a sample's age; they depend on many other factors as well, including sample size and type, and details of the measurement technique. They are inherent in all the dating methods discussed in this book. But they can be quantified by well-known statistical techniques, and they are always reported along with the dates, providing a good sense of how reliable a particular age determination really is. Sometimes they are referred to as "errors," but that term implies a mistake or problem, and I prefer (and will use throughout the book) the term *uncertainty* because that is really what they are.

Libby was awarded the 1960 Nobel Prize in Chemistry in recognition of his development of radiocarbon dating. The citation from one of those who nominated him read: "Seldom has a single discovery in chemistry had such an impact on the thinking of so many fields of human endeavor. Seldom has a single discovery generated such wide public interest." That research in nuclear chemistry involving a rare radioactive isotope should generate any public interest at all might at first seem remarkable. But, as we have seen, radiocarbon dating is especially useful for dating events in human history, and Libby's work struck a chord with the public. Who would not be intrigued by the discovery of a way to measure the age of Egyptian kings, or to date man's first foray into North America?

The agreement between the early radiocarbon ages and those determined from historical or other reliable evidence—especially for those first few samples included in the "curve of knowns" (see figure 7 on page 66)—seemed almost too good to be true. But, as more samples were analyzed during the 1950s, and as the measurement uncertainties decreased because of improved experimental methods, some disturbing trends showed up. Occasional "fliers"—dates that seemed to be completely wrong—were not the problem; these could usually be explained by human error such as sample mislabeling, or perhaps by contamination of the sample with material of a different age (I will come back to the problem of contamination later in this chapter). What was troubling, however, was that there seemed to be a consistent trend of samples being "too young" by up to several hundred years. This discrepancy could only be discerned in cases where there was firm archaeological evidence for the true age, but there were enough of those to cause concern. Was there a simple explanation, or was radiocarbon dating turning out to be unreliable?

The standard reaction among those who debunk dating based on radioactivity is to seize on such apparent discrepancies and declare that *all* ages measured using these techniques are nothing but fabrications. The more rational response, however, is to ask where the problem might lie. And the first step in doing that is to look at the assumptions that underlie the dating method.

In the case of radiocarbon dating, determining the age of a sample is, in principle, quite straightforward. The first step is to measure its carbon-14 content accurately, and the second is to plug this measured value into the radioactive decay equation and calculate the age. The equation is the same one that Ernest Rutherford formulated from his observations of the systematic decay of the radioactive gas radon. Although I have avoided using equations in this book, the general decay equation (see appendix C), adapted for carbon-14, is reproduced below because it is so important. The radioactive decay equation is actually not difficult to understand, and, if you make the effort, it should help to clarify the nature of radioactive decay and its application to dating. Figures 5 and 7 on pages 56 and 66 both show the equation in graphical form (as solid curved lines). It is written as.

$$^{14}C = (^{14}C)_o e^{-\lambda t}$$

What the equation says is that the amount of carbon-14 in a sample (this is the measured value, represented by the term on the left-hand side) is equal to the amount of the isotope present at time zero, $(^{14}C)_o$, times the expression $e^{-\lambda t}$. The symbol e is the standard representation for a mathematical constant (the number 2.71828 . . . , which is used as the base for natural logarithms), and λ is another constant that characterizes the rate at which carbon-14 decays (it is directly related to the half-life). The age to be calculated is represented by the symbol t for time.

The two major assumptions in radiocarbon dating involve the two terms $(^{14}C)_o$ and λ on the right-hand side of the equation. If those two parameters are known accurately, then the only quantity in the equation that *isn't* known is *t,* the age of the sample, which can then be calculated quite easily. So the question has to be asked, Do we know those two parameters accurately enough for the method to work properly?

When Libby embarked on his development of radiocarbon dating, the half-life of carbon-14 (and therefore the constant λ) was not well known. Values ranging from 5,000 years to over 25,000 years had been reported, and several of these estimates had large uncertainties. Libby and several of his colleagues decided to make their own measurement for use in the dating work, and they averaged the result they got with what they considered to be the most accurate of the previous determinations. In this way they came up with a half-life of 5,568 years, with an uncertainty of just plus or minus 30 years.

But, as it turned out later, this result—which became known as the "Libby half-life"—was slightly off. More recent determinations place the half-life of carbon-14 at 5,730 years; this is the accepted value today. The difference is small, just under 3 percent. For the most part, it had no effect on Libby's work because the measurement uncertainties were at least this large for most of the early radiocarbon dates. On the other hand, it did mean that all ages calculated using the Libby half-life were, in a statistical sense, systematically on the low side of their true ages. As more and more data accumulated, this discrepancy became obvious, but it was also easy to correct later on, simply by recalculating the dates using the newer, more accurate half-life value.

The parameter $(^{14}C)_0$ in the decay equation, however, is more problematic. It represents the carbon-14 content of the sample material at the time the plant or animal died, and it obviously can't be measured directly. Ernie Anderson's measurement of the "contemporary assay" showed that the radiocarbon content of living things is the same everywhere on Earth today, and it seemed reasonable to assume, at least as a first approximation, that this value also characterized organic carbon in the past. The agreement illustrated in Arnold and Libby's "curve of knowns" reinforced this conclusion, and suggested that it was valid at least over the past few thousand years. But beyond the reach of the historical record, there seemed to be no obvious way to test this assumption. Or was there?

One of the scientists whom Jim Arnold instructed in his minicourses on radiocarbon dating was a brilliant Austrian chemist named Hans Suess, who had been invited to visit the University of Chicago in 1949. Suess had heard about Libby's carbon-14 work and was interested in its potential as a dating tool. While he was in Chicago, he took every opportunity to learn as much as he could about the new technique.

Suess never did return permanently to Austria. He ended up staying in the United States, and went on to an illustrious research and academic career. Some years after leaving Chicago, he became a professor of chemistry at the University of California at San Diego, where he had the reputation of being the quintessential absent-minded professor. That trait was apparently evident quite early on, because Arnold says that while most of the "students" who visited Libby's lab to learn about radiocarbon dating took copious notes and were extremely attentive, Suess just wandered in, listened for a while, and went away again. But this casual approach, like his veneer of absent-mindedness, disguised a sharply perceptive mind. Suess had quickly assimilated the basics. He didn't need to take notes about Arnold's procedures because he already had his own ideas about how to improve the method. Suess soon left Chicago for Washington, D.C., where the U.S. Geological Survey (USGS) had asked him to set up a radiocarbon dating laboratory. Over the next few years, he and his colleagues carried out a large number of key carbon-14 analyses, concentrating especially on the chronology of ice age glaciation in North America.

Suess's most important contribution to the rapidly developing field, however, came after he left the Geological Survey and moved to San Diego. There he began collaborating with researchers at the Laboratory of Tree-Ring Research, part of the University of Arizona in Tucson. Dendrochronology—the science of tree-ring counting—was already a valuable tool in archaeology. With care, by counting back from the present, each annual growth ring can be assigned to an exact calendar year. In their "curve of knowns," Arnold and Libby had included dates from the inner portions of two ancient trees, illustrating that the radiocarbon and tree-ring dates agreed (see figure 7 on page 66). That result demonstrated that wood in a growing tree ceases to exchange carbon-14 with the environment once it is formed (otherwise the dates would not have agreed). Each growth band captures the carbon-14 signature of the atmosphere during the year it grows, and then becomes a kind of sealed time capsule.

Researchers at the Tree-Ring Laboratory had been able to push dendrochronology back far beyond the time span of living trees by patching together overlapping growth-band sequences. Because of year-to-year changes in temperature and precipitation, the appearance of tree rings varies, especially their thickness. Two or three years of drought produce two or three years of thin rings; a wet year produces a spurt of growth and a thick tree ring. Over time, a unique pattern comes to characterize all the trees in a region. By painstakingly matching up overlapping patterns between living and dead trees, and then between increasingly older dead trees, researchers had been able to put together a continuous record that could be traced back several thousand years. Wood from an archaeological site could often be dated simply by comparing its tree-ring pattern with this master record.

Suess understood the potential of "calibrating" radiocarbon ages by using tree rings. Because each ring records the carbon-14 content of atmospheric CO_2 in the year it grows, he could test Libby's assumption that the carbon-14 content of living matter—the parameter $(^{14}C)_0$ in the decay equation—had not changed over time by measuring radiocarbon in rings of accurately known ages. Some of the wood investigated at the Tree-Ring Laboratory—notably from no-longer-living bristlecone pines (see figure 8) that grow at high elevations in the mountains of California—was as much as 7,000 years old. If Suess could measure the "radiocarbon age" of such samples at closely spaced intervals from the present back to 7,000 years, he would have a much more detailed curve of knowns than the one Arnold and Libby had published. If the tree-ring and radiocarbon results agreed, fine; if there were discrepancies, it

Figure 8. A dead bristlecone pine, named the "Colossal Ghost" by its photographer, Leonard Miller. Not only do these trees have very long lives, but, in the arid climate where they grow, their wood resists decay long after death. Scientists can compile a long, continuous tree-ring record by matching ring patterns. Photograph copyright Leonard Miller.

would imply that $(^{14}C)_0$ had varied—and also, the data could be used to adjust and correct dates measured for unknown samples.

Actually, when Suess began his work on tree rings, he was aware that there might be differences between his radiocarbon dates and those obtained by counting rings, because Hessel de Vries, a Dutch scientist, had already completed similar work on wood from European trees. His results, published in 1958 and 1959, showed that there indeed were significant discrepancies, and, importantly, that the offsets were not constant (as would occur if they were due only to an incorrect half-life). De Vries attributed this finding to natural variations in the amount of carbon-14 in the Earth's atmosphere in the past. The discrepancies were immediately dubbed the "de Vries effect."

However, de Vries died tragically in 1959 at a young age, while Suess continued working on radiocarbon dating of tree rings. In 1961, he published data that ranged back to approximately 3,000 years, and, later, in 1969, he extended the range to 7,000 years. Today, he is the scientist most closely associated with the early calibration data. By measuring closely spaced samples through 7,000 years of history, he was able to identify and analyze both short- and long-term trends in the data.

Suess's results, like those of de Vries, showed that the offset between radiocarbon and tree-ring dates changes in systematic ways. There are gradual, long-term (thousands of years) variations, but also an abundance of shorter-term (hundreds of years) wiggles superimposed on those longer variations (see figure 9). De Vries had been right. The only explanation for these patterns was that the amount of carbon-14 in the ancient atmosphere must have varied with time. In retrospect, it is easy to say: Of course! Why would anyone expect the radiocarbon content of the atmosphere to remain constant over thousands of years? Libby himself probably recognized that his assumption would only be true to a first approximation. The work of de Vries and Suess recalls an aphorism attributed to Enrico Fermi, the Italian physicist who was instrumental in building the world's first nuclear reactor at the University of Chicago: "If you make a measurement and get what you expect, you have made a measurement. It you don't get what you expect, you've made a discovery."

The wiggles in atmospheric carbon-14 discovered from the tree-ring data were definitely an important discovery, and, as we will see later in this chapter, they have implications that extend beyond radiocarbon dating. But their immediate significance was their demonstration that one of the fundamental assumptions of the method was not really valid. All was not lost, however, because the same data that revealed the discrepancies could be used to correct them. By 1969, Suess's "calibration curve" was already quite detailed, and more tree-ring samples were being analyzed every day. With good coverage for a particular age range—like that in figure 9—the true age of a sample could be determined directly from the calibration curve. It's easy to see that, for intervals over which the curve rises smoothly and rapidly, such as between about 3,320 and 3,420 years (calendar age),

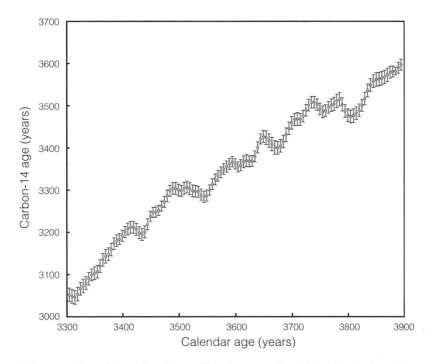

Figure 9. Part of the radiocarbon calibration curve, in which calendar (actual) ages, based on tree-ring counts, are plotted against ages calculated from carbon-14 measurements. Because carbon-14 in the atmosphere has varied in the past, the two do not agree precisely. Uncertainties in the carbon-14 data are shown by the vertical bars at each data point; the consistency of the measurements is remarkable, allowing even small variations in past atmospheric carbon-14 to be identified. The short-term "Suess wiggles" are very evident in this example, but the gradual, longer-term variations are not, because of the short time period shown. Data for this graph are from the most recent radiocarbon calibration, by Paula J. Reimer et al., *Radiocarbon* 46 (2004): 1029–58.

this approach is quite sensitive. For other regions of the chart, such as that between 3,750 and 3,850 years, ambiguities arise. In fact, several "correct" ages may be permissible because of the wiggles.

All this manipulation can seem pretty confusing if you are encountering the details of radiocarbon dating for the first time. In a vague way, it seems to the uninitiated that radiocarbon daters are somehow fudging their results. But, in reality, the procedure is a straightforward result of experimentation and observation. The carbon-14 data for tree rings from the western United States show exactly the same wiggles and variations at exactly the same times as those from Europe. The agreement is impressive—not only do the data come from laboratories using different measurement techniques, but also the annual growth-ring patterns in the trees reflect local climatic variations, and are therefore different from region to region. In spite of these variations, however, there is worldwide consistency in the calibration data, indicating that the tree-ring ages are accurate and that the radiocarbon measurements closely track global variations in the amount of carbon-14 in the ancient atmosphere, even when they were quite small. (A very small offset occurs between the Northern and Southern Hemispheres because of the details of carbon-14 production and mixing in the atmosphere, but this phenomenon is well understood and does not affect the overall conclusion.) In effect, the calibration curve provides a value for the term $(^{14}C)_0$ in the decay equation. This value is not constant, as was initially assumed, but as long as its variability through time is known, it can be taken into account and will not affect the accuracy of the dates. To prevent confusion, in the scientific literature, the results of carbon-14 analyses are reported, by convention, as uncorrected "radiocarbon ages" (the vertical axis in figure 9). These are calculated from the laboratory measurements using agreed-upon values both for the half-life and for the present-day carbon-14 content, $(^{14}C)_0$. The true age of the sample can then be read from the appropriate portion of the calibration curve. And, again to avoid confusion, the calendar (true) ages are always referred to 1950. Thus, a sample dated as 3,000 years old from the calibration curve was 3,000 years old in 1950; in 2000, it was 3,050 years old, and so on.

Quite aside from the implications of the calibration curve for dating, it is also a window through which to examine other phenomena that have affected the Earth in the past. When Suess showed his data to

Libby and pointed out that there seemed to be some regularity to the wiggles, Libby reportedly said, "If this is true, then the radiocarbon values should be a most interesting geophysical parameter." That has certainly turned out to be the case. As de Vries first suspected, the wiggles—and also the longer-term variations—are caused by past changes in the carbon-14 content of the atmosphere. But, in a sense, that is only a symptom. The question is, What caused these variations?

As is true for many natural phenomena, the pattern of changes revealed in the calibration curve does not have a single cause; instead, it is the result of several different processes. One of these has to do with the way carbon cycles through the various "reservoirs" in which it resides on Earth, such as the atmosphere, the oceans, and living organic matter. But more important is the strength of the cosmic ray bombardment that produces carbon-14 in the first place. To a large extent, both the wiggles and the longer-term variations are a kind of fossil record of this interaction—the greater the intensity of cosmic ray bombardment, the more carbon-14 is produced, and vice versa.

Only a small fraction of the cosmic rays traveling through space toward our planet actually make it to the atmosphere, because the Earth's magnetic field acts as a shield, deflecting most of the particles away. And the magnetic field is constantly changing; direct measurements show, for example, that its strength has decreased by about 10 percent since the eighteenth century, and much greater changes have occurred over longer time scales. When the field increases, fewer cosmic ray particles can penetrate through; when it decreases, more make it to the atmosphere. Such changes therefore affect the production of carbon-14, and most researchers believe magnetic field changes are behind the long-term variations evident in the calibration curve. These variations, however, are too gradual to explain the short-term wiggles. The consensus is that these are related to changes in the sun's activity.

It might seem strange that the sun can affect carbon-14 production in the Earth's atmosphere. But the sun produces its own magnetic field, which extends out into space far beyond the Earth. When the sun is active, this field strengthens, and fewer cosmic ray particles make it to the Earth's atmosphere, leading to lower carbon-14 production. This has been verified in an ingenious way. Times of high solar activity are marked by an increase in visible sunspots, and vice versa. Because sunspots have intrigued observers for millennia, there is an almost continuous record of their occurrences since at least 2,000 years ago, when Chinese scholars began recording them. To the extent that this record can be compared with the radiocarbon calibration curve, the two agree: times in the past when numerous sunspots were noted correspond to lower carbon-14. The active sun generated a stronger magnetic field, more effectively shielding the Earth from cosmic rays. And there is another possible correlation as well. Because variations in the sun's activity affect the amount of solar energy reaching the Earth, they can affect climate. Thus the radiocarbon calibration curve may also harbor information about past climate change.

Hans Suess was well aware that his data extending back to 7,000 years held important clues to past interactions between cosmic rays and the Earth's atmosphere. He put a lot of effort into trying to understand the significance of the small-scale variations in the curve, and they soon became known as "Suess wiggles." But he was bemused by the reaction of many of his fellow scientists. He wrote later that it was of great interest to him to "[understand] the psychological causes that led the great majority of investigators to deny, for many years, the existence of regular deviations of the carbon-14 values, the so-called 'wiggles,' from a smooth line." He was not the first to be perplexed by the tendency of some researchers to resist new ideas, even in the face of compelling experimental evidence. Even scientists, he thought, prefer straight lines and regularity to the messy pattern of wiggles his data revealed.

I hope it's clear from the last few pages that most of the radiocarbon dates that appeared to be "wrong" based on archaeological evidence—especially in the early days of the technique—were not wrong at all. The measurements were accurate, and the age calculations appropriate. What made them appear "wrong" is, first, that, because of past variations in its production rate in the atmosphere, the amount of carbon-14 in living matter has not always been the same as it is today, and, second, that Libby's initial half-life determination was slightly in error. The calibration curve, which has now been extended back tens of thousands of years, even beyond the time scale accessible through tree-ring studies, takes care of those effects. With accurate knowledge of carbon-14's half-life, and of carbon-14 variations in the past, radiocarbon measurements give a true measure of a sample's age.

Keeping in mind that the calibration curve is continually being improved, as are the instruments for making carbon-14 measurements, let's take a look at how early successes quickly put radiocarbon dating on the map in two important research areas: glacial geology, and the entry of humans into the Americas. Although more recent work has sharpened the details, the research by Libby and his colleagues, and by a few other early practitioners of radiocarbon dating, laid out a remarkably accurate framework on which all subsequent investigations have been based.

During the first half of the twentieth century, glacial geology was an especially popular and well-studied aspect of the earth sciences in North America and Europe. The reason is not difficult to understand. Throughout the northern United States and most of Canada, and in Scandinavia, Great Britain, and the northern fringes of Europe, the surface landscape has been heavily influenced by the great glaciers of the ice age. Signs of their presence are everywhere if you know where to look. You don't even have to go out into the countryside—even in Central Park, in the heart of Manhattan, you can see the surface polish and scratches left on rocky outcrops by the sandpaper-like action of flowing, grit-filled glaciers. In other places, huge boulders, carried by the ice and left stranded when it melted, sit in farmers' fields, too heavy to move. Mounds of rocky rubble scraped up by glaciers and dumped along their borders have created gently undulating topography in the same regions. The fertile fields of the American Midwest were developed on the finest grains of that rubble, winnowed and transported by the wind. As the glaciers melted back, lakes developed in low-lying areas along their margins, depositing layer upon layer of characteristic glacial sediments. All these features were mapped out in great detail by geologists, beginning in the nineteenth century. By using the cardinal rule of stratigraphy, that young material always covers or cuts through older material, geologists were able to work out the relative chronology for some of the events of the ice age. But they had to rely on crude estimates of how fast natural processes occurred—how long it took soil to develop on glacial rubble, for example, or how fast ice flowed or melted back—as a guide to the actual time scales. And, because the glacial deposits are not always continuous, it was difficult to correlate from one region to another even over reasonably short distances, and virtually impossible to determine whether deposits in Europe and North America had been laid down at the same time. This made it hard to know whether the glacial deposits reflected global or simply regional changes in climate, and also to work out the causes of the glaciation. But nearly all estimates of the timing of glacial activity put at least some of it in the geologically recent past, thousands or tens of thousands of years ago. Thus glacial geology was a field ripe for radiocarbon dating.

Richard Foster Flint, the lone geologist on Libby's advisory committee, was the most prominent glacial geologist of his day, a professor at Yale and an eloquent speaker. Much of his research involved mapping and interpreting glacial deposits—he virtually single-handedly put together the first glacial map of Connecticut. Flint was acutely aware of the importance of accurate chronology for this work, and to him radiocarbon dating seemed to be a godsend. He could easily have kept the Libby laboratory busy full-time with samples related to the ice age. However, archaeologists were also clamoring for dates, and, in the first list of radiocarbon ages in *Science* in 1951, the majority of results reported by Arnold and Libby were archaeological. Nevertheless, a few of the samples had been chosen for their glaciological significance, most on Flint's recommendation. They included peat and mud rich in organic material from several sites in Europe, and also a number of samples from North America. Most important was a group of samples from a location known as the Two Creeks Forest Bed.

Two Creeks has acquired worldwide fame among radiocarbon chronologists. The site lies in Wisconsin, along the western shore of Lake Michigan, and long before radiocarbon dating became a reality it had provided a fascinating and almost unparalleled snapshot of ice age processes in action. By the end of the nineteenth century, geologists had established that during the most recent ice age there had been, at a minimum, four major advances and retreats of glaciers across northern North America. The last of these they named the "Wisconsin" glaciation because some of its most striking effects are seen in that state. (To preserve and highlight this legacy, in 1971 the National Park Service and various local organizations established the Ice Age National Scientific Reserve of Wisconsin, which includes a one-thousand-mile-long Ice Age National Scenic Trail that wends its way across the state, passing through and over the glacially sculpted landscape.) The glacial sediments at the Two Creeks locality had been assigned to the Wisconsin glaciation, and geologists had established that they revealed a complex series of events probably representing the very last gasps of the Wisconsin. The most striking feature of this site was the preserved remnants of a forest that had been literally toppled over by the last advance of the great ice sheets. The event was, so to speak, frozen in time. As the ice melted away, a huge lake formed—a precursor of present-day Lake Michigan—and deposited mud and silt over the destroyed forest, sealing it off from further disturbance. It was an ideal target for radiocarbon dating: an age for this site would establish the time of the last significant glacial episode in this part of the country, and possibly North America as a whole.

To give you an inkling of the glacial record preserved in the Two Creeks sediments, figure 10 shows an idealized cross-section through a bluff along the Lake Michigan shoreline in the area. Like the sketch James Hutton's friend made of the rocks in Jedburgh, Scotland (see figure 2 on page 10), this seemingly simple picture provides a wealth of information about past events. First, at the bottom is a layer of loose, rub-blelike sediment characteristic of the material that glaciers scoop up and push along, and then leave behind when they melt, commonly referred to as "glacial till" by geologists. Its presence is an unequivocal sign of a period of glaciation. Overlying the till is a layer of much finer material, clay and sand, exhibiting its own layers and evidently deposited in a lake that covered the site after the glaciers receded. Above the lake sediments sits a soil horizon (labeled "forest bed debris" in figure 10), dark and peaty, containing bits of still-obvious pine needles, pinecones, and other organic material. Occasionally there is a broken tree trunk, standing vertically, its roots still anchored in the soil layer. These remnants record a time when the lake receded, vegetation flourished, and trees grew. But again the region was flooded, as revealed by another layer of clay and sand covering the soil and engulfing the tree trunks. On top of that is a further layer of till, signifying another glacial advance. It still contains some of the tree trunks sheared off by the glaciers, nearly all of them lined up in the local direction of ice movement. Still higher up on the bluff is yet another layer of lake sediments, deposited as the glaciers melted, again flooding the land.

Most of us wouldn't give a second thought to a sequence of layers like this exposed on a hillside. But, to a geologist, every layer is brimming with information about the end of the Wisconsin glacial period, which is why the Two Creeks locality is so important. To help elaborate on the story told by the sediments, countless samples have been taken from the various layers and carefully examined for pollen grains, plant remains, and shells. Together they provide a comprehensive picture of the local vegetation and give clues about the climate. Of most interest, as far as radiocarbon research is concerned, are the soil layer and the broken trees that are the remains of a forest flooded by rising lake waters and bull-dozed by the advancing glacial ice.

Preservation at the Two Creeks site is so good that an amazingly detailed picture has been constructed of the forest ecosystem. The wood is mostly spruce and hemlock, typical of northern forests today, and some of

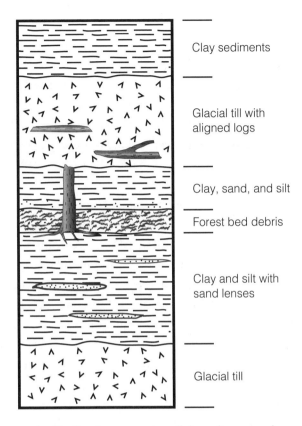

Clay sediments

Glacial till with
aligned logs

Clay, sand, and silt

Forest bed debris

Clay and silt with
sand lenses

Glacial till

Figure 10. An idealized cross-section of the sedimentary layers at the
Two Creeks locality. Beds of glacial till record two glacial episodes, the
second of which sheared off parts of trees in a forest that had existed for
at least sixty years. Libby's radiocarbon dates on material from the "forest bed" showed that the last glaciation of the area occurred 11,400 years
ago. This drawing is based on information in a monograph by C.E.
Prouty for the 1960 field excursion of the Michigan Basin Geological
Society.

the trunks even retain remnants of bracket fungus, as well as holes from
at least two different species of bark beetles. Mosses, plants, and mollusks
that lived in the forest have also been identified. Tree-ring counts show
that the average age of the trees was about sixty years when the forest was
flooded and cut down by the advancing glacier. By selecting a specific

block of rings for radiocarbon analysis, it might be possible to date the last surge of Wisconsin glaciation to within a few tens of years.

Prior to carbon-14 dating, geologists had estimated that the ice advance recorded by the sheared-off Two Creeks trees dated to somewhere between 20,000 and 25,000 years ago. It is not surprising, then, that the Arnold and Libby radiocarbon age for wood from the soil layer caused quite a stir: it was only 11,400 years. This was completely unexpected. Until then, no one had imagined that ice had reached so far south in North America so recently.

The Two Creeks result was a turning point in understanding glacial chronology, and for some it ranked as the most important date in the whole of the 1951 Arnold and Libby paper. According to Jim Arnold, Richard Foster Flint accepted the date without great surprise; perhaps he already had an inkling that the much older age conventionally associated with the site was incorrect. For some other researchers, though, it was a controversial result. But, as more ages were measured for glacial deposits, and as several new laboratories began churning out radiocarbon dates, it became clear that the Two Creeks age was not an anomaly. The height of the Wisconsin glaciation, the time of maximum ice, had indeed occurred at around 20,000 years ago, when glaciers reached far south of the Great Lakes, covering much of present-day Illinois, Indiana, and Ohio. Sometime after this, they began to melt back, but it was a slow process. Not until nearly 10,000 years ago did they begin their rapid and irrevocable retreat in the face of a warming climate. In between, especially along the margins of the ice sheet in the northern United States, there were local advances and retreats in response to small changes in climate conditions. During cold periods, lobes of ice would reach down valleys like the fingers of a hand, only to draw back again when temperatures rose. The Two Creeks age dated the very last of these episodes.

Since Arnold and Libby's work, the geography of ice margins and the advances and retreats of localized ice lobes have been meticulously documented using evidence similar to that found at Two Creeks. Most of the important localities have been radiocarbon dated, providing a detailed chronology of ice movement. The ages have been crucial for correlating events at widely separated locations, and they have shown that most swings in climate, even quite small ones, were simultaneous both in Europe and North America.

Given its importance, the Two Creeks Forest Bed itself has been redated by other laboratories. With the benefit of the calibration curve, which had not even been thought of when Arnold and Libby published their first dates, the later analyses gave a slightly older age. The difference, however, was only a few hundred years, and it did not change the conclusion that retreat of the Wisconsin ice sheet occurred much more recently than had once been thought.

Libby and his group had to rely on Richard Foster Flint and other geologists to advise them about important samples for glacial chronology, but they did have a connection with archaeology, however tenuous, through Jim Arnold's knowledge of Egyptian history. Partly for that reason, many of the early archaeological samples that they dated came from Egypt. Libby also asked Arnold to be a liaison between the laboratory and archaeologists, and sent him to a number of archaeological conferences. Arnold took this responsibility seriously. In addition to attending meetings, in the summer of 1949—after working flat out for months in the laboratory—he took his "vacation" by going to an archaeological field camp.

The camp was at a permanent Field Museum site in western New Mexico, and it provided Arnold valuable insight into the intricacies of archaeological sampling. It also introduced him to New World archaeology. Libby himself had an abiding interest in using radiocarbon dating to work out the timing of human migration into the Americas, and because several members of the advisory committee, including the chairman, Fred Johnson, were also deeply involved in New World archaeology, it was natural that this soon became a focus of dating activity in the Chicago laboratory. It was a focus that paid off handsomely, because the data Libby and his colleagues collected during their first few

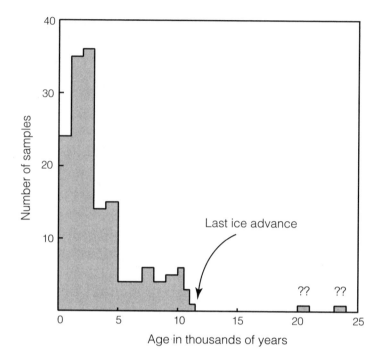

Figure 11. For his 1960 Nobel lecture, Libby plotted a bar graph
like this one of all existing radiocarbon dates for North American
archaeological sites. He emphasized the link between the end of the
Pleistocene Ice Age and the spread of early people by showing the
time of the last ice advance as determined from his date for the Two
Creeks site (11,400 years).

years of work provided the initial chronological framework for American archaeology.

By the time of his Nobel Prize lecture in 1960, Libby was able to present a striking bar graph that included all radiocarbon dates then available for North American archaeological sites (see figure 11). An important feature of the graph is the very abrupt cutoff in ages near 11,000 years ago; only two sites gave older dates, and both were thought to be questionable, possibly because of contamination. (Although I have said little about contamination, it is a serious issue for radiocarbon dating. For old samples that contain very little carbon-14, addition of even a small amount of "modern" carbon can be disastrous. Incorporation of just a few minute root hairs from living plants that penetrate into charcoal at an archaeological site, for example, or fungus growing on a museum sample, will make the measured age much too young. "Dead" carbon from fossil fuel–based products such as paraffin or oil can result in an age that is far too old. Many apparently aberrant radiocarbon dates, especially in the early days before the severity of the problem was realized, were the result of just such contamination.)

Recall that the Two Creeks glacial deposit, dated by Arnold and Libby at 11,400 years, was thought to record the last significant advance of ice on the North American continent before the glaciers melted away entirely. To Libby and many others, the absence of archaeological sites older than this was no coincidence. Extensive peopling of North America, they concluded, followed the retreat of the great continental-scale ice sheets of the Pleistocene Ice Age. Libby marked the Two Creeks age on his bar graph to emphasize this conclusion.

Even today, countless analyses later and almost half a century after Libby first showed this graph, it is still the case that most radiocarbon dates for humans in the Americas are less than 11,400 years, although there are significant exceptions. There is also still a vigorous debate about the extent to which glaciers of the Pleistocene Ice Age aided or hindered migration from Asia into and across the Americas. But at the beginning of the radiocarbon work, the picture was not nearly so clear. One of Libby's first attempts to analyze an ancient North American site using radiocarbon—a site that he thought would date to the earliest habitation of the New World—produced an age that was much younger than the time of the glaciers. It was a locality characterized by artifacts linking it to what archaeologists refer to as the Folsom culture.

Folsom is a small town in northeastern New Mexico. In 1926–27, arrowhead-like stone points were found there, mixed together with bones from a now-extinct type of bison, a discovery that caused great excitement because it placed humans in New Mexico during the last glacial period, when bison were abundant. Beyond that general observation, however, there was no way to date the site. Eventually, additional "Folsom" sites were discovered in other regions, all characterized by the same distinctive stone points. A few of these were in places that could be correlated with specific glacial deposits, which, through a fairly tenuous line of reasoning, were thought to be between 10,000 and 25,000 years old. Most workers favored the older end of the range.

Arnold and Libby included a charcoal sample linked to the Folsom culture in their first published list of radiocarbon dates. The result was a surprise: 4,283 ± 250 years. This was clearly much younger than any of the earlier estimates suggested, and, if the date held up, it would mean that what appeared to be one of the oldest Native American cultures was actually quite recent. Although they were confident about their analysis procedures, Libby and Arnold were suspicious of the result and wondered if the sample had been contaminated with young carbon, or if there was some other difficulty they were unaware of. In the end it turned out to be a classic case of improper sampling, and an example of the importance of careful field documentation. When the charcoal was collected in 1933 (it had been stored away from then until the analysis), it appeared to be lying within a soil layer that contained both animal bones and the distinctive Folsom stone points. But the unexpectedly young age prompted reexamination of the site, and it was discovered that the charcoal came from a channel that cut into and through older layers. Although it appeared to be at the same level as the bones and stone points, it was actually much younger. Once this problem was recognized and new samples from this and other sites were analyzed, it became clear that the most reliable Folsom ages fell in the range of 10,000 to 11,000 B.P. (before the present).

However, it was also discovered that Folsom sites are not the oldest evidence for humans in North America. At some localities, slightly different varieties of stone hunting points occur; initially it was thought that these were simply regional variations, or perhaps weapons used for hunting different types of game. But, in some places—notably at Clovis, New Mexico—they appear in layers that lie *beneath* the typical Folsom points.

This indicated that they were older, and soon archaeologists began to distinguish between Clovis and Folsom cultures. Obviously, Clovis sites became another target for radiocarbon dating, and the results confirmed their antiquity. Clovis sites consistently gave dates that were a few hundred years older than those characterized by the Folsom artifacts, and there seemed to be little or no overlap between the two cultures.

With these results, the radiocarbon dates of both glacial deposits and archaeological sites in North America seemed to be painting a consistent picture. As the last severe glaciation of the Pleistocene Ice Age waned, early people spread into the United States. Clovis people were the first widespread hunters, making distinctive stone points for their weapons and hunting large game such as mammoths. Within a few hundred years, however, a new culture appeared, making smaller and finer stone points and apparently taking over from its Clovis predecessors as the dominant hunters in North America.

The notion that the Clovis people were the first important culture to populate the Americas has been prevalent from the time of the early radiocarbon analyses until quite recently, and the few ages that seemed to indicate there were even older inhabitants, before that key date of about 11,400 years ago, were viewed with suspicion. But it is now evident that some sites really are much older. One is Monte Verde in southern Chile, where repeated and carefully scrutinized radiocarbon results provide strong evidence for human occupation at approximately 12,500 years B.P. Monte Verde is almost as far south as one can go in the Americas, and, if humans migrated into North America from Siberia and spread southward, as seems to be the case, it is obvious that at least some people crossed over long before 12,500 years ago. That, however, is something of a puzzle, because within the United States and Canada almost all ages are much younger. Some archaeologists think that the earliest migrants could have taken coastal routes, avoiding the extensive inland glaciers that then existed and making their way south to ice-free coastal Oregon and California before spreading inland or farther south. Firm evidence for such migrations is lacking, but it has been argued that because sea level was much lower at the time, any sites in low-lying coastal regions have since been submerged under rising seas.

Radiocarbon dating, then, brought the earlier fuzzy chronology of North American prehistory into sharp focus. By providing accurate ages for the deposits of the last great ice sheets and the campfires of paleoindians, it enabled geologists and archaeologists to map out the movements of both, and to investigate the interrelationships between the two. It has also done much more than can be explored here to increase our understanding of the changing flora, fauna, and climate of North America (and elsewhere) from the time of the glaciers to the present. None of that could have been accomplished without the detailed chronological framework provided by carbon-14.

CHAPTER FIVE

Getting the Lead Out

By a curious coincidence, in the late 1940s and early 1950s, just as Libby, Arnold, and Anderson were developing radiocarbon dating as a way to measure samples that are just thousands of years old, there was a parallel project underway to develop methods capable of dating samples from the other end of the time spectrum: near the time of the Earth's formation, billions of years before the present. Not only was this work taking place almost simultaneously, it was going on literally right next door at the University of Chicago.

During the Second World War, many of America's best chemists and physicists were involved in the Manhattan Project, a dispersed, nationwide effort to develop the atomic bomb. One of the centers for this work was at the University of Chicago. Robert Hutchins, who was then the university chancellor, saw firsthand the powerful synergy this project generated by bringing together bright people of diverse backgrounds to tackle a common problem. He wanted to keep this spirit of creativity alive, so, when the war ended, he founded the Institute for Nuclear Studies (now known as the Enrico Fermi Institute) at the university as a center of excellence for nuclear research. It was also a way to keep Manhattan Project scientists at Chicago, and in this Hutchins was supremely successful; not only did many of the researchers stay, but some of their colleagues were attracted to Chicago from other universities. Although the personnel have changed with time, the institute and its affiliated departments have generated a steady stream of high-quality research ever since.

The scientific work on the Manhattan Project had been intense and highly rewarding for many of the participants, but it was focused on the single goal of developing nuclear weapons. With the war over, many scientists who had been involved in that effort were having second thoughts about the forces they had unleashed, and began to turn their newly acquired knowledge of atomic processes away from military problems and toward pure science. Several of those who ended up at the University of Chicago chose to work on problems in earth and solar system science. Without really realizing it, in doing so they founded an entirely new field of research: isotope geochemistry. Today that field is pervasive. It is now a rare exception to find an issue of an earth science journal that doesn't include papers dealing with isotopes or nuclear processes.

The group of scientists working in geochemistry at Chicago was an unprecedented accumulation of talent. It included, among many others, Bill Libby; Harold Urey, who, like Libby, was a Nobel Prize winner (for his discovery, long before the war, of an isotope of hydrogen); and the chemist Harrison Brown. One of Urey's most famous accomplishments at Chicago was the development of a method to determine the temperature of ocean water in the past by measuring isotopes of oxygen in fossil shells. This technique is now in use in dozens of laboratories around the world, and it has become crucially important for investigating the Earth's past climate in the context of present-day global warming. Harrison Brown had wide interests in using chemistry to understand the Earth, and he also had a knack for gathering together really good people and getting them involved in interesting geochemical projects (it was Brown who invited Hans Suess, the Austrian chemist of "Suess wiggle" fame, to Chicago).

Libby, Urey, and Brown interacted frequently, sometimes shared laboratory facilities, and were well versed in each other's research projects. They also attracted bright students and postdoctoral fellows to work with them, many of whom later moved on to prestigious positions around the country and were largely responsible for the rapid growth of isotope studies in the earth sciences. Although all three would eventually turn their eyes westward and move to California, while they were together in Chicago there was pure excitement in the air. The field was young, and the possibilities for making important discoveries seemed almost unlimited. Radiocarbon dating is an obvious example. When two talented graduate students—George Tilton and Clair Patterson—showed up on Harrison Brown's doorstep, he gave them an assignment that typified the prevailing atmosphere: develop a method for measuring the ages of ancient rocks.

At the time, little was known about the earliest parts of the Earth's history. Abundant fossils—which, as we will see later in this book, geologists had used to work out sequences of events in our planet's past—only appear in sedimentary rocks beginning with the Cambrian period of the geological time scale, which we now know began 542 million years ago (see appendix A). There was a fairly good understanding of the *relative* time scale from the Cambrian to the present, and even a rough outline of numerical dates in this interval, based on uranium-lead dating. But of Precambrian time, the time before fossils, very little was known. Vast areas of the continents are covered with ancient, contorted rocks that contain no fossils whatsoever; these are the great metamorphic "shields," such as the Canadian Shield of North America. Nobody knew how old most of these rocks were.

The search for the oldest rocks of the Earth's crust, which is really what Brown's assignment for Tilton and Patterson was all about, had always been linked to the bigger question, How old is the Earth? After all, the most ancient rock to be found would provide a minimum age for the Earth itself. By the time Tilton and Patterson began their work, the age of the oldest known rocks had been pushed back to a few billion years. Much of the credit for that work must go to a British geologist, Arthur Holmes, who devoted most of his career to developing an accurate geological time scale.

Holmes was a young student at Imperial College in London in the early years of the twentieth century, when Rutherford, Soddy, and others were uncovering the nature of radioactivity and the atom. He was studying physics and was fascinated by the new discoveries, but after taking a course in geology he was hooked. He decided—against the advice of his physics professors—to switch subjects. Holmes never looked back, becoming one of the most highly respected geologists of his day.

Although he was still a student, it seemed obvious to Holmes that radioactivity had a major role to play in geology. He had read the work of Bertrand Boltwood, Ernest Rutherford's friend at Yale University, who had established beyond reasonable doubt that the end product of uranium decay is lead. Isotopes were still unknown, but in 1907 Boltwood published ages for several uranium-rich rocks simply by assuming that all the lead they contained had come from the decay of uranium. The dates ranged from 400 million to 2.2 billion years. But there was a problem: the half-life for uranium decay, a crucial parameter in the age calculation, was not known with any certainty. Boltwood had to estimate it by using data from Rutherford's experiments with radium, one of the intermediate products in the chain of radioactive decays between uranium and lead. It was not a very satisfactory solution.

A small diversion is in order here to examine the principles behind Boltwood's "uranium-lead" method for determining the ages of rocks. As we saw in chapter 3, for radiocarbon dating, the approach is to determine how much of the original carbon-14 has decayed away, using the assumption that the original radiocarbon content was the same as that in living material today. For uranium-lead dating, though (and for all the other dating methods examined in this book), the situation is somewhat different. For these techniques, it is virtually impossible to know the sample's original content of the radioactive isotope used for dating. So, instead of determining how much of the isotope has decayed away, the important measurement is of the amount of daughter isotope that has accumulated—the product of the radioactive decay. By plugging this value into the radioactive decay equation, together with the present-day content of the radioactive parent (another value that can be measured directly), an age can be calculated. The important thing to remember is that the key measurement in uranium-lead dating is of the lead atoms that have accumulated through decay.

To return to our story, Arthur Holmes published his first uranium-lead date in 1911, as part of an "original research" requirement for his undergraduate degree. He was still only twenty-one, but the work brought him rapid recognition. Holmes had planned his experiment carefully, choosing a rock from Norway that, on the basis of its geological setting, was believed to date from the Devonian period. (Modern research shows that the Devonian period lasted from 416 to 359 million years ago; during this time, the first trees and insects appear in the fossil record.)

The sample contained several different types of uranium-rich minerals, which Holmes carefully separated; in principle, each of these could be used for a separate age determination and thus serve as a check on the others. Holmes followed Boltwood's procedure, extracting tiny quantities of lead and uranium from his mineral samples and weighing them. This required great skill in analytical chemistry, because each sample had to be dissolved in acids, then run through a series of steps to separate out pure lead and uranium compounds, free of any contaminants. With each step there was the possibility of loss of material simply through handling, and, in hindsight, the accuracy of many of the early age determinations—even though they were still quite crude by today's standards—seems remarkable.

Holmes had to perfect his techniques through experience and by making mistakes along the way. There were some false starts, but he was determined, and anyway he really wanted that degree. Finally he got the procedures to work, and calculated the age of his Norwegian rock: 370 million years. Because he was deeply immersed in the study of geology, he realized immediately that this result had importance beyond simply showing that the uranium-lead method was useful for dating rocks: his measurement gave an absolute age for the Devonian period. The entire geological time scale was then still a relative scale, based on the order in which certain fossil organisms appear in rocks worldwide. Rutherford was a physicist and Boltwood a chemist, and both viewed their rock dating work primarily as a demonstration that radioactivity could be used to determine the ages of things. In contrast, Holmes saw uranium-lead dating as a way to quantify and refine the relative geological time scale. It was a task that would occupy him for much of the rest of his career.

Holmes published his data for the Norwegian rock only four years after Boltwood's earlier work, but in the meantime the half-life of uranium had been determined more accurately. That gave him confidence in the results, and it also prompted him to recalculate Boltwood's results using the new uranium half-life value. When he did that, the ages turned out to be significantly younger—Boltwood's oldest sample became 1.64 instead of 2.2 billion years old. To the extent that independent geological information was available for Boltwood's samples, Holmes also noted where each of them fit in the relative geological time scale, enabling him to establish the beginnings of a true absolute chronology for geological history.

Thus, in little more than a decade, the prevailing view about the Earth's age had shifted from Lord Kelvin's 20 million years to more than 1.5 billion years. The new dates based on radioactivity were accepted by most scientists, and they underscored the intuitive belief of many geologists, from the time of Hutton onward, that extremely long time periods were necessary to explain many geological phenomena. There was, however, a residue of skepticism. As late as 1924, a prominent geologist in the United States Geological Survey, F. W. Clarke, opined that various lines of evidence actually pointed to an age for the Earth of between 50 and 150 million years. "The high values found by radioactive measurements," he wrote, "are therefore to be suspected until the discrepancies shall have been explained."

By the time Harrison Brown asked his two graduate students to find a way to measure the ages of ancient rocks accurately, nearly forty years had passed since Holmes's and Boltwood's first uranium-lead analyses.

Much had been learned in the interim about radioactive decay. Among other things, isotopes had been discovered, and it had been found that there are two different isotopes of uranium and four of lead. The simple uranium-to-lead dating idea suddenly became much more complicated. Simply measuring the uranium and lead contents of a sample wasn't enough to give an accurate age; instead, it would be necessary to measure the quantities of each of the isotopes separately. Brown wanted Tilton and Patterson to do this by employing a device that had been used widely for isotope measurements during the Manhattan Project, a mass spectrometer. As the name implies, this instrument quantifies samples in terms of mass—put in some uranium, and it tells you how much is uranium-235 and how much is uranium-238. Similarly, it could separately measure each of the four isotopes of lead. That sounded very simple, and in principle it is, but, as the two graduate students were to find out, making the measurements accurately and reliably is a complex and difficult task.

Some dating work had already been done using mass spectrometers, but most of it had been carried out on samples that contained large amounts of uranium—especially uranium ores—because these were the only kinds of rocks in which enough lead had been produced by radioactive decay to measure by the prevailing techniques. However, Harrison Brown had learned about work being carried out at the U.S. Geological Survey on a mineral rich in zirconium, called zircon. For Brown, zircon had a number of desirable characteristics: it contains quite a bit of uranium; it is widely distributed in rocks of the Earth's crust; and, because of its crystal structure, it does not incorporate lead when it forms. This latter feature meant that virtually all the lead in an ancient zircon would be the product of radioactive decay—a crucial condition for accurate uranium-lead dating. Because nearly every outcrop of granite in the world contains crystals of zircon, Brown knew that developing a method to date these crystals would mean that age determinations would no longer be restricted to rare uranium ores or uranium-rich samples. It would be possible to date almost any part of the Earth's crust, and to systematically investigate questions about how the continents have evolved over time.

Brown had another motive as well. Like others, he was keen to use lead isotopes to find the age of the Earth. And he wanted to do this not just by finding and dating the Earth's oldest rock. All the equations describing the decay of uranium to the various isotopes of lead had been worked out, and he knew that in principle it should be possible to date the whole Earth in the same way as a zircon crystal, by plugging values for the various parameters into the equations. But there was one key set of necessary values that posed a problem. Unlike a zircon crystal, the Earth contained lead when it was formed, inherited from the material that made it up. Because the age equations take into account only the lead produced directly by radioactive decay, it would be necessary, somehow, to determine the isotopic composition of the lead that was present initially before an accurate age could be calculated.

Brown thought that the initial composition could be found by measuring meteorites, because the Earth had been formed from meteorite-like material. However, the measurement techniques would have to be perfected before the measurement itself could be attempted. He told Patterson that once he had learned to analyze lead isotopes in zircon crystals, doing the same for an iron meteorite, and then calculating the Earth's age, would be "duck soup." And, Brown told him, "You'll be famous."

That all sounded very promising to Patterson, but putting it into practice turned out to be a very long and involved process. Duck soup it certainly wasn't. Brown had divided the dating task between the two students; once they had separated the zircon crystals, Tilton was to measure the uranium isotopes, and Patterson the lead. Because the zircon crystals were tiny, the amounts of both elements in their samples would be small. This was particularly true for lead; the quantity available would be only about one one-thousandth the amount that anyone else had ever measured before.

Just as Arnold and Libby had analyzed samples with known ages as the first step in developing radiocarbon dating, so Brown wanted Tilton and Patterson to start with zircon crystals from an already-dated rock as a test of the method. The best Brown could do was to find a granite sample that was associated with—and presumably the same age as—a dated uranium ore. But, unlike the radiocarbon results, when the first measurements were made, they did not agree with the known age. After examining the data closely, the researchers concluded that the uranium content measured by Tilton was correct. It was the lead analyses that were suspect—not only did Patterson's measurements give far higher concentrations of lead than expected, they also indicated that the lead isotopes were present in the wrong proportions. Thinking that there had been a problem in the analysis, Patterson repeated the measurements. But the results were the same.

This was discouraging. It had already taken about a year to get the analytical techniques to the point where Tilton and Patterson could accurately measure the small amounts of uranium and lead in their samples, and now there seemed to be a major problem. The only reasonable explanation, the two students realized, was contamination. Somehow, extraneous lead from the environment was getting into their samples and producing the spurious results. The question was, Where was it coming from?

The discovery that his samples were contaminated would, quite literally, change Patterson's life. For most of the rest of his career, his energies would be focused on ways to reduce contamination so that ever-smaller samples could be analyzed accurately. Patterson's work would make him quite famous among geochemists because, in many cases, his were the only measurements they trusted. Researchers came to his laboratory from around the world to find out how he did it. And, in a good example of how pure research often has unexpected outcomes, society as a whole benefited from Patterson's efforts. The harmful health effects of lead were already well known, but it was largely his research that revealed the ubiquitous presence of lead in the modern environment and, eventually, prompted action to reduce it.

As he worked on the ancient zircons, Patterson found that there was lead in everything. It was in the chemicals he used to dissolve the crystals, in the beakers he used in the laboratory, even in the dust particles floating around in the air. The absolute amounts were small, but, in relation to the quantities in his tiny zircon crystals, they were far too large—large enough to produce dates that were completely wrong. Patterson would find and eliminate one source of contamination only to discover that there was yet another that was just as serious. In an interview in 1995, shortly before he died, Patterson asked the interviewer if she remembered the cartoon character Pigpen, from the *Charlie Brown* comic strips. Pigpen, he reminded her, is the one always portrayed with stuff flying off him in all directions. "That," said Patterson, "is what people look like with respect to lead. Everyone. The lead from your hair, when you walk into a super-clean laboratory like mine [Patterson was talking about his 1990s laboratory here], will contaminate the whole damn laboratory."

It took Patterson literally years of work to reduce contamination to the point where he got the "right" answer for zircon grains separated from granite of known age. He had to learn, by trial and error, where the main sources of contamination lay. He had to make his own chemical reagents by distilling the components, often repeatedly, to get rid of the lead they contained. Laboratory ware that came in contact with his samples—such as beakers—had to be boiled in acid. Dust in the laboratory had to be reduced or eliminated. It was an impressive accomplishment, and it became a major part of Patterson's 1951 PhD thesis. But he had not forgotten his conversations with Brown about the age of the Earth. It was something Patterson very much wanted to pursue now that he had perfected the analytical methods and finished his PhD work. He asked Brown if he could stay at Chicago as a postdoctoral researcher to work on the problem, and Brown concurred. Brown knew the question was important, and he also knew that, with the experience Patterson had gained in analyzing zircons, he was better equipped than anyone to address the question.

As already pointed out, the main difficulty in determining the Earth's age was that nobody had worked out a clever way to estimate the lead isotope abundances of the Earth when it formed. Because of the unusual situation in the uranium-lead decay system—two different uranium isotopes that decay to two different lead isotopes—it is possible to manipulate the equations in such a way that ages can be calculated based solely on the ratio of the two lead isotopes. This is different from other dating methods, which generally require measurements of both parent and daughter isotopes. To calculate the Earth's age, all Patterson would have to know was two numbers: the ratio of lead-206 to lead-207 in the Earth when it formed, and that same ratio today. The difficulty came down to finding samples that could be used to establish these values.

Patterson was not the first person to tackle this problem. Several other researchers, including Arthur Holmes, had tried, using analyses of ancient lead ores made by Alfred Nier, of the University of Minnesota, as a best estimate for the Earth's initial lead isotopic composition. The rationale was that these ores contained no uranium, and therefore their lead isotopes had not changed because of radioactive decay since they were formed. Nier was a highly respected physicist who had vastly improved the earliest versions of mass spectrometers, building new instruments that were capable of very precise isotope measurements. His results were widely agreed to be accurate; the problem, however, was in the choice of samples. The lead ores were very old, but they didn't date from the time of the Earth's formation, and therefore they couldn't really be used to determine its initial isotopic composition. The ages calculated using these values fell, for the most part, between 3.0 and 3.5 billion years. This pushed the age of the Earth back almost another 2 billion years, but it was still far from being an accurate value.

Harrison Brown thought that a much better estimate could be obtained by measuring iron meteorites. By the 1950s, many researchers realized that meteorites are just one part of the spectrum of materials—from tiny dust grains in space to the planets and the sun itself—that make up the solar system. All must have formed at about the same time, and from the same precursor matter. If that were so, meteorites and the Earth would have had the same initial lead isotope composition. The iron meteorites, like the lead ores analyzed by Nier, contain vanishingly small amounts of uranium, so radioactive decay has not altered their lead isotopes. Measuring them today should give the "primordial" value for the Earth.

For Patterson's work, Brown got several chunks of an iron meteorite named Canyon Diablo, which had crashed to Earth 50,000 years ago and created a spectacular three-quarters-of-a-mile-wide crater about forty miles west of Flagstaff, Arizona. Although most of the meteorite had vaporized on impact, enough survived that collectors have picked up an estimated thirty tons of material, and there is still lots left. Patterson needed only a few grams for his analysis, so there was no shortage of material. Iron meteorites are composed almost entirely of metallic iron, which does not occur naturally on Earth. Evidently this strange material fascinated indigenous people as much as it does scientists today; pieces of Canyon Diablo have been found together with other artifacts at several archaeological sites in the region.

Brown's hunch that iron meteorites would contain "primordial" lead was a good one. Patterson's analyses showed that the Canyon Diablo sample had the lowest abundances ever measured of both lead-206 and lead-207, the two isotopes produced by the decay of uranium, proving it was very ancient. The results were published in the journal *Physical Review* in 1953, and, at a conference that same year, Patterson showed that if he plugged these values into the appropriate equations, he calculated the Earth's age to lie between 4.51 and 4.56 billion years. Ask any geologist today about the age of the Earth, and he will give you a number that falls within that range.

Just as his mentor, Harrison Brown, had predicted, determining the age of the Earth with such precision and rigor made Patterson famous, at least among geologists and other scientists, if not the public. Although he would soon turn his attention to other problems—mostly still involving lead—he was not yet quite finished with dating the planet. There was still the nagging question of whether lead isotopes in the Canyon Diablo lead meteorite really did represent the Earth's "primordial" lead. It was just an assumption, and, although it seemed reasonable, it was not proven.

To test the assumption, Patterson analyzed several more meteorites, together with a sample of Pacific Ocean sediment. The latter was meant to represent, in terms of its lead isotopes, an average sample of the continents surrounding the Pacific, and therefore an approximation of the average lead isotope composition of the Earth's crust today. The meteorites were selected to include several different varieties of the so-called "stony" meteorites, which have much higher uranium contents than iron meteorites like Canyon Diablo. If all these samples—including the Earth—had started out with the same lead as Canyon Diablo, then each would now have a distinctive isotopic composition that was a function of its uranium content. Such a result would indicate a direct connection between the Earth and the meteorites, and it would confirm that the lead isotopic composition of the Canyon Diablo meteorite was a reasonable choice for the Earth's primordial lead.

When Patterson plotted his data on the appropriate graph, the points defined a straight line (see figure 12). Like Arnold and Libby's "curve of knowns," this figure, published in the geochemical journal *Geochimica et Cosmochimica Acta* in 1956, has become a classic diagram in the earth sciences. Canyon Diablo, with very low lead isotope ratios, established one end of the line, while the other meteorites, all containing significant amounts of uranium, plotted at much higher values. The Pacific Ocean sediments plotted partway along the line, close to values for two of the stony meteorites. A straight line on this diagram is the graphical representation of the uranium-lead decay equation for a series of samples with the same age; the slope of the line is a measure of that age. The fact that all Patterson's samples fell along such a line meant that they had the same age. When he calculated the age from the slope of the line, Patterson got 4.55 billion years. Furthermore, the lead isotope ratios in the Canyon Diablo iron meteorite fell at one end of the line, indicating that all the samples had started out (4.55 billion years ago) with that same lead isotope composition, but, over the intervening time, they had accumulated lead-206 and lead-207 through radioactive decay of the uranium they contained (the curved lines in figure 12 show

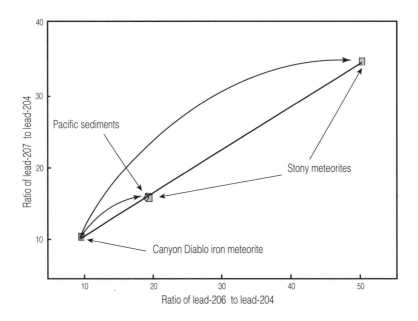

Figure 12. Patterson's age of the Earth diagram is one of the most famous diagrams in the earth sciences. It shows the lead isotopic composition of several meteorites and a terrestrial sample, plotted as ratios of lead isotopes produced by radioactive decay of uranium (lead-206 and lead-207) to an unchanging isotope, lead-204. (Two of the "stony" meteorites Patterson analyzed have very similar lead isotope ratios, and the values are also close to that for the Pacific sediments. At the scale of this drawing, the three data points overlap one another.) Although the graph may not look like much, the fact that several meteorites and a sample of Pacific sediments all plot along a straight line indicates that they share a common age. This property arises from the lead isotope equations (see page 249), which, when expressed in terms of the ratios on the axes of Patterson's graph, become the equation of a straight line for all samples of the same age. The slope of the straight line gives the age—4.55 billion years in the case of Patterson's data. The straight-line relationship also shows that the lead in all the samples analyzed by Patterson started out with the same isotopic composition as the iron meteorite Canyon Diablo 4.55 billion years ago, and that, because of uranium decay, each has evolved over time (following the curved lines) to its present composition. Samples with the most uranium have evolved furthest from their original composition. Based on data in C. Patterson, *Geochimica et Cosmochimica Acta* 10 (1956): 230–37.

how their compositions have changed over time; Canyon Diablo, with no uranium, has remained the same).

Based on his precise lead isotope measurements and the close fit of his samples to a straight line, Patterson assigned an uncertainty of just plus or minus 70 million years to the 4.55-billion-year date. He had established beyond reasonable doubt a precise age for the Earth. But his result was doubly important because the common age and common initial lead isotope composition of both meteorites and terrestrial material definitively linked—for the first time—the Earth with diverse other solar system objects.

In the years since Patterson's age of the Earth was published, much effort has gone into refining or correcting his result. Patterson said later that there were a lot of people who "worked their hearts out to prove I was wrong." He wasn't being paranoid—he was just pointing out that scientists are always probing and testing important results. He had, he said, the best critics in the world because they really wanted to find a flaw in his work. But they didn't. The key to Patterson's success was his ability to reduce contamination to a level of insignificance, a level that would not affect even the "primordial" lead isotope compositions in iron meteorites. It is easy to think that this might be a simple task, but you must realize that no earthly sample has a composition anything like that of the lead in the Canyon Diablo meteorite. Allowing the tiniest amount of lead from the Earth to creep into Patterson's samples would have ruined the analysis. He was able to make more accurate measurements of the true isotopic compositions of his samples than anyone else because he understood how important it was to prevent contamination.

In 1951, just as Patterson was getting started on his age of the Earth work, his mentor, Harrison Brown, was appointed professor at the California Institute of Technology in Pasadena, California. Patterson soon followed him west. At Caltech he had better laboratory facilities and ever-improving control over lead contamination, and he carried out the chemical processing of samples for his age of the Earth study there (however, the processed samples were then carefully transported back to

Chicago for mass spectrometer measurements). And although he continued to work on the ages of meteorites and the rocks of the Earth's crust,
at Caltech he also took advantage of his hard-earned expertise and began
to examine lead isotopes in a wide range of materials that had never before been properly analyzed. Among these were ocean sediments and
river water. Both have very low lead concentrations, and Patterson's ability to minimize contamination was a fundamental factor in the research.

The lead in deep-sea sediments comes from at least two sources. Some
of it resides in the grains of clay that are washed into the sea from the
continents, and some of it is deposited into the sediments directly from
seawater. This dissolved lead is brought into the ocean by rivers, and,
other things being equal, the input from rivers should be balanced by the
output to the sediments, so that the overall lead concentration in seawater stays approximately constant. Most natural systems tend toward such
a balance. But Patterson's analyses showed that there was a huge discrepancy: the lead content of river water was far higher, by about a hundred times, than was required to balance the amount in sediments. As
Fermi said, when your measurements give you a value you don't expect,
you've made a discovery. The river water analyses, Patterson knew,
measured the present-day lead input, while the sediments reflected the
average deposition in the past. He had discovered that much more lead
is flowing into the oceans today than had in earlier times.

Patterson investigated further and found that the lead content of shallow ocean water is much greater than that of deep water. In snow and ice
from polar regions, he found that the near-surface layers, representing
the past few hundred years of accumulation, contain much, much more
lead than snow from earlier times. Initially, there was skepticism about
his results, because nobody else could reproduce the measurements.
Other labs found far higher lead contents in almost everything they
analyzed, including ocean water and deep-snow layers. But Patterson
knew that these higher lead contents were due to contamination. He
pointed out that even the laboratory water used by some of these other
researchers to make up their analytical chemicals contained a thousand

times higher lead concentrations than his polar snow and ice samples. It was no wonder others couldn't replicate his results.

But where was all this lead in river water, snow, and his competitors' samples coming from? Patterson realized that it had to be from human activity, specifically industrial sources, because all his research suggested that such lead had begun to appear in just the past few hundred years. Furthermore, his work showed that the rate of increase of lead concentrations in the environment had risen dramatically in recent decades. The primary culprit, it turned out, was an antiknock chemical that had been added to gasoline since the 1920s: tetraethyl lead. The lead from every tankful of gasoline containing this additive went straight out the exhaust pipe and into the atmosphere, to be breathed by humans, incorporated into snowflakes, or to settle out on the ground and be washed into rivers and oceans. It even snuck insidiously into chemical laboratories. In the 1970s, not long before lead additives were banned from gasoline in the United States, one of Patterson's Caltech colleagues, a scientist who built his own "clean lab" to analyze moon samples returned by the Apollo missions, was fond of pointing out that there was more lead in one cubic centimeter of Pasadena air than in many of his Apollo samples. It all came from gasoline.

If lead were inert, Patterson's findings would have been interesting, but not particularly startling. However, even in small doses, lead is a poison. The gasoline additive tetraethyl lead is highly toxic; factory workers have died from contact with small amounts of the substance, a fact that its manufacturers tried hard to keep quiet. The great irony of Patterson's discovery is that his work on deep-sea sediments, which started the entire saga, had been funded by the petroleum industry. Harrison Brown had convinced the oil companies that lead isotopes would be a good "fingerprint" for tracing sediment layers from one region to another, something they were deeply interested in. They were quite happy to spend money on research that might give them a new and useful exploration tool, but they were less than thrilled when Patterson's work uncovered the health hazards posed by everyday use of leaded gasoline in automobiles. His funding ceased abruptly. According to Patterson, his petroleum industry funders also quietly pressured some of his other sources of research grants, notably the U.S. Atomic Energy Commission, to cut his funding.

Fortunately, these machinations were not successful, and Patterson was able to continue his investigations. Among other things, he discovered that we have almost a thousand times more lead in our bones than did people living a few thousand years ago, a factor so far above "natural" concentrations that it approaches the level of lead poisoning. Patterson's work, which he always claimed was initiated more by pure curiosity about the natural world than by environmental concerns, was a major factor, quite probably *the* major impetus, in the push to reduce lead contamination of the environment through such measures as banning tetraethyl lead from gasoline, and eliminating lead-containing solder from food cans. The concern continues, as illustrated by the 2007 recall of children's toys in the United States because they were coated with lead-containing paint.

The research on environmental lead has had wide-ranging consequences, but, in scientific circles, it was Patterson's accurate determination of the Earth's age, and the connection he drew between the Earth and meteorites, that brought him most recognition. More than half a century has passed since that work, and there have been major advances in the precision with which lead isotopes can be measured; the half-life of uranium has been measured more accurately; and the control of lead contamination in research laboratories has improved greatly. Nevertheless, the basic conclusions Patterson reached have not changed.

What has changed, though, is our understanding of some of the details surrounding the very earliest history of the Earth and the solar system. That understanding got a quite significant push forward through the confluence of several serendipitous events. Just after midnight on February 8, 1969, at a time when laboratories across the United States (and, to some extent, in other countries as well) were gearing up to analyze samples they expected to be brought back from the moon by the *Apollo 11* astronauts, a huge fireball burst across the skies of Mexico and the southwestern United States. It was a meteorite, a large one, weighing ten tons or more before much of it burned up in the atmosphere. As it slammed through the air, it began to fragment, its pieces falling to earth over a broad band of territory near the village of Pueblito de Allende in Chihuahua, Mexico. It is standard procedure to name meteorites after the village or town closest to their place of fall, so this one became known as the Allende meteorite.

Meteoriticists, as those with a special interest in meteorites are known, will hop on an airplane at the drop of a hat if there is a chance of finding a new sample of one of these unusual rocks from outer space. When it became clear that Allende was both very large and one of the rarest varieties of meteorite, hordes of geologists, museum curators, and other collectors descended on Pueblito de Allende to pick up the pieces. In all, they recovered more than two tons of material, and, although the region has been pretty thoroughly combed over, there is still more to find for those with sharp eyes and patience. Because samples were so abundant and Allende was an uncommon type of meteorite, it was quickly probed and analyzed in laboratories the world over. What has given this particular meteorite special importance is that it contains large numbers of small, rocky inclusions that have turned out to be the oldest known material in the solar system (see figure 13).

Not long before Allende crashed into the ground in Mexico, a graduate student at Yale University named Larry Grossman submitted his PhD thesis, a theoretical treatment of what might have happened during the early days of the solar system. The general view at the time was that the sun, Earth, and planets had formed when a cloud of gas and dust—similar to the clouds astronomers observe in other parts of the universe today—collapsed in on itself. Most of the material from the cloud ended up in a very dense and very hot central blob in which nuclear fusion reactions began, forming our sun. For his PhD research, Grossman attempted to answer the question, What happened as the leftover and still very hot gas around the infant sun began to cool?

Figure 13. A fragment of the Allende meteorite showing numerous calcium-aluminum-rich inclusions (CAIs), the small white or light-colored objects visible on the cut surface. This specimen is about 4.25 inches across. Photograph courtesy of Daniel Ball, Center for Meteorite Studies, Arizona State University, Tempe.

Grossman approached this problem by setting up a whole series of equations formulated to show what kinds of minerals would condense out as this hot gas cooled down. He found that many of the earliest minerals to form would be highly enriched in the elements calcium and aluminum, and that they would appear when the gas was still extremely hot. His calculations also showed that, as the temperature continued to drop, many other solid materials would condense, some of them minerals that are quite familiar on Earth. However, it is those first, high-temperature minerals that relate Grossman's thesis work to the Allende meteorite.

Had Allende not fallen when it did, Grossman's thesis might have languished in relative obscurity, read only by the few people interested in the problem he had investigated. The work itself was comprehensive and a significant improvement on the few prior attempts to tackle the same question, but it was, after all, a theoretical solution. The scenario he sketched out was plausible, but there was no guarantee it had really occurred. What gave Grossman's thesis research instant fame, however, was the fact that the small, light-colored inclusions in Allende, most of them no more than a few millimeters—less than an eighth of an inch—across (see figure 13), were highly enriched in calcium and aluminum. This was just what his calculations had shown for the first solid materials to form in the solar system. Many of the individual minerals in the inclusions were the same ones he had predicted, too.

Inevitably, the inclusions are known as CAIs, standing for calcium-aluminum-rich inclusions. They have been studied intensively since Allende's fall, partly because of their abundance, but mainly because of the possibility that they are messengers from the earliest moments of the solar system's history, largely unchanged since the time of their formation. They have been analyzed for their mineral content, their chemical compositions, and, of course, their isotopic characteristics and their ages. Some of the more unusual ones have been given names by the researchers who studied them. Most of the CAIs have quite high uranium contents and can be dated accurately using uranium-lead dating, and their measured ages corroborate Grossman's conclusions—the CAIs are very ancient. They do indeed seem to be the first solid materials formed in our solar system.

A recent paper in *Science* shows just how old these remarkable objects are, and it also illustrates the precision with which uranium-lead dating can now be done. The paper reported on a study of two CAIs (not from the Allende meteorite in this case, but from another meteorite of a similar type), giving their age, as determined by uranium-lead dating, as 4.5672 billion years. The authors quote an uncertainty in their measurements of only plus or minus 600 *thousand* years. Imagine an analysis so precise that the timing of an event that happened 4.5 billion years ago can be nailed down to within a few hundred thousand years! So as not to have quite so many digits after the decimal point, the age was reported as 4,567.2 million years. It could also be written as 4,567,200,000 years. That is slightly older than Patterson's age of the Earth, but not much.

Most scientists regard Patterson's 4.55-billion-year age for the Earth as dating the time when our planet had grown to about its present size. The few refinements that have been made to the original age have changed the value slightly, making it just a few tens of millions of years younger than the oldest CAI dates. This means that the building of planets in the solar system took place very quickly in geological terms. From the appearance of the first small bits of solid matter—like the CAIs in the Allende meteorite—to an essentially full-sized Earth required only tens of millions of years. Although there were no witnesses to document this process, there are sophisticated computer models of how it might have occurred, and they are in agreement with the uranium-lead dates. They predict that the early formed CAIs and other small objects would continually collide with one another as they orbited the young sun, sometimes sticking together and sometimes breaking apart. Those that stuck together grew larger and larger, sweeping up all the material they encountered along their orbits. Eventually, through a random process, the situation approached the current state: the sun surrounded by a few large planets, as well as some remaining smaller bodies such as the asteroids and the moons of the larger planets. The models indicate that the time required for all of this to happen would be measured in millions to a few tens of millions of years.

Although we now know quite a bit about the chronology of this earliest segment of the Earth's and the solar system's history, there are still significant gaps in our knowledge about the early days of the Earth's existence as a fully fledged planet. This, remember, is what Harrison Brown originally wanted Patterson and George Tilton to do: to find a way to date the oldest parts of the Earth's crust. Because these ancient rocks occur in the very old Precambrian shields that form the cores of all continents, they invariably have a complex history. Most of them have been through the geological mill; they have been swirled and folded and contorted to the point where it may be difficult to know what the original material was. "Like peanut butter and jelly," one of my geological colleagues would say of rocks like the stretched and visibly layered Precambrian gneisses, and that is indeed a good description. Many of these ancient bits of the continents have been rafted around the Earth's surface by plate tectonics (the all-encompassing geological theory that explains how mountains, earthquakes, volcanoes, and many other features are caused by the constant moving about of great blocks of the Earth's surface that smash into, slide past, or plunge under one another); buried deeply under great mountain ranges that rivaled the Alps or the Himalayas in size; exhumed again; covered over by tropical seas; and scoured by ice age glaciers. Often their ages provide information only about the last time they were buried and heated to the "peanut butter and jelly" stage, close to their melting point. But sometimes, as we will see shortly, it is also possible to get a glimpse of an earlier history.

In the years following Patterson's determination of the Earth's age, the search for the world's oldest rocks was concentrated in areas where the Precambrian shields had already been well studied and there was reason to believe that very old rocks existed: Australia, Canada, and the northern United States; South Africa; and Greenland. Especially in the 1970s and 1980s, after a burst of innovation in analytical methods ensured that smaller and smaller samples could be measured more and more accurately, the age of the oldest known rocks climbed rapidly, from around 3.0 billion years initially to 3.2 billion years, then to 3.4, 3.5, 3.6 . . . , and so on until the present, when the record stands at just over 4.0 billion years. In this quest, being able to find and recognize the most ancient parts of the Earth's crust has been at least as important as improved analytical techniques. And beyond 4 billion years the record is still blank—well, almost blank; we will see that there are ghosts of earlier rocks, but no actual specimens. What little we know of the

Earth's first half billion years comes from just a few tiny crystals that would barely cover the bottom of a small thimble.

In fact, almost half of the Earth's history is very sparsely represented by actual samples. Less than 15 percent of the present continental area is occupied by rocks that are more than 2.5 billion years old, and the further you push back in time, the rarer samples become. For samples dating to 3.8 billion years or older, the total exposed area dwindles to just a few hundred square miles. And determining the age of these ancient metamorphic rocks—which themselves were formed by deep burial and heating of even older rocks—is often problematic. There is always the question of whether a date provides information about the time of their metamorphism or the age of their precursors—or whether the metamorphism has hopelessly muddled the results, making the date meaningless.

Fortunately, the mineral zircon has proven to be a geochronologist's dream come true. Not only does it incorporate quite a bit of uranium and very little lead when it forms, it is also very stable over a wide range of conditions. It can be heated to almost 900 degrees Celsius (more than 1,600 degrees Fahrenheit) and remain relatively unaltered, a crucial attribute for dating ancient metamorphic rocks. At low temperatures, when weathering turns some minerals to clay, zircon crystals survive untouched. Scoop up sand from almost any beach, and some of the crystals you hold in your hand will be zircons.

Because of these favorable characteristics, zircon has become the sample of choice for dating the Earth's oldest rocks. Other methods can be useful in some cases, but for overall applicability, uranium-lead dating of zircon crystals is by far the most widespread approach. Laboratories around the world have developed the capability to analyze just a few small crystals of this complex mineral and extract a date. In some cases, tens or even hundreds of pounds of rock are crushed and processed to obtain the crystals, which are sorted for analysis on the basis of size, color, and other physical attributes (see figure 14). Hard-won experience shows that, even among crystals from the same rock, some give more accurate ages than others, and visible characteristics can often

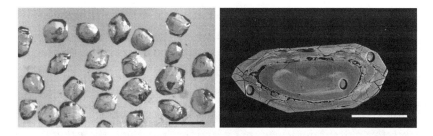

Figure 14. Different views of zircon crystals. Those on the left, viewed through an optical microscope, were separated from an igneous rock from eastern Australia for use as a dating standard. They are clear and homogeneous, and have been dated at 417 million years by uranium-lead dating. The scale bar is 200 micrometers long, or about 0.008 inch. The photograph on the right shows a cross-section of a single zircon crystal from Manitoba, Canada, that was embedded in epoxy, polished flat, and examined using a scanning electron microscope. The scale bar is 100 micrometers long, or about 0.004 inch. This crystal was separated from a metamorphic rock and has a complicated internal structure with a distinct core and rim. The small, craterlike holes are analysis spots from ion microprobe measurements (described in more detail in chapter 9). The measurements show that the core of the crystal is just under 2.5 billion years old, while the rim is much younger, 1.8 billion years old. Photos courtesy of Sandra Kamo and Mike Hamilton, Jack Satterly Geochronology Laboratory, University of Toronto.

be a clue to picking out the "best" grains. Because zircon crystals sometimes grow additional outer layers during metamorphism—meaning that even a single crystal may contain portions that formed at widely separated times—elaborate methods have been developed to gently abrade away the outer regions of the grains before analysis, leaving only the inner cores that presumably retain the original age of the rock. And, following in Patterson's pioneering footsteps, most laboratories have now developed techniques for stringently controlling contamination, so that, in the best cases, single zircon crystals, or sometimes even fragments of a single crystal, can be dated.

Tilton and Patterson (and most researchers since) had to dissolve their zircon crystals in strong acids to prepare them for analysis. Because zircon crystals are so stable, even this is not easy, and the necessary handling can introduce contamination. But, in a technological development that has revolutionized uranium-lead dating, a new instrument called an ion microprobe (described in more detail in chapter 9) has made it possible not only to avoid dissolving zircon crystals, but to perform *multiple* analyses on a single grain. Such analyses are done in situ, directly on the grain (see figure 14), and they repeatedly show that zircon crystals from ancient metamorphic rocks are very complex. Quite often, the multiple analyses result in multiple ages for different parts of the same crystal.

Ion microprobe analyses were used to date the current record-holding 4.0-plus-billion-year-old rocks that are the oldest yet found. Zircon crystals were extracted from samples of the Acasta Gneiss, a metamorphic rock complex in the northwestern corner of the Canadian Shield, the swath of ancient rocks that forms the core of the North American continent. Sam Bowring, a geologist at the Massachusetts Institute of Technology (MIT), was the first to discover the great age of the Acasta Gneiss when, in 1989, he and several colleagues dated zircons from it to 3.96 billion years. The rock outcrops are complicated, with crude layering and peanut-butter-and-jelly swirling (see figure 15); adjacent rock layers can give quite different ages, and the zircons are also complex. They have been affected by multiple metamorphic events, and ion probe analyses show that different parts of the grains sometimes date from events separated by as much as hundreds of millions of years. Detailed, careful mapping of the rock outcrops was necessary to find the oldest parts of this complex, and only the oldest zircons from the oldest layers gave the 3.96-billion-year age.

The lure of the Earth's oldest rocks exerts a magnetic pull on geologists, and since Bowring's work was first published the Acasta Gneiss locality has been visited by scores of researchers eager to examine, sample, and analyze these remnants from a time so distant that it challenges the imagination. Chronology using ion microprobe measurements on zircon crystals has been the mainstay of that work, and the age

Figure 15. An outcrop of the Acasta Gneiss of northern Canada, the oldest intact rocks yet found on Earth. Note the complex, swirled texture of this rock, which has endured multiple metamorphic episodes. Photo courtesy of Dr. Yuichiro Ueno, Tokyo Institute of Technology.

of the oldest zircons has now been pushed back to slightly more than 4 billion years. The picture that has emerged, in its broadest outline, is that the oldest rocks were originally a type of granite that crystallized just over 4 billion years ago. Since then, these rocks have been subjected repeatedly to metamorphism, which transformed them into the gneisses they are today. Those metamorphic events included deep burial, heating, and mixing with younger rocks during massive collisions between ancient continents. There are hints in the data that, when the original granites formed, they might have incorporated zircons from even older rocks, which has spurred further detailed field studies in an attempt to locate these precursors. None have yet been found, however.

The Acasta Gneiss is the oldest known rock complex on Earth that you can actually visit and stand on and pick apart with a geologist's hammer. But I mentioned earlier that there are "ghosts" of even more ancient material. Once again, it is the remarkable stability of zircons that allows us to peer so far back into Earth's history—but, in this case, what is important is not their ability to withstand the high temperatures and pressures of metamorphism, but their stability at the much lower temperatures of surface weathering.

One of the most common varieties of sedimentary rock formed from the weathering debris of continents is sandstone. It is made up mainly of tiny grains of quartz, because, among all minerals in rocks, quartz is both very common and highly resistant to degradation. But nearly all sandstone also contains zircons, weathered out of the same rocks as the quartz, and geologists have long studied these crystals as tracers to track the origin of the sandstone components. Dating the zircons and comparing the results with the ages of possible precursor materials often makes it possible to work out the ultimate source of the sediments. So why not separate and date zircons from the oldest sandstones you can find? They would indeed be the ghosts of their parent rocks, and even if those parental materials no longer exist, the zircons would retain information about their ages. Perhaps that would be a way to break the 4.0-billion-year barrier and probe even further back into the Earth's history.

It is just such a possibility that persuaded researchers at the Australian National University in Canberra to begin the tedious work of separating and analyzing large numbers of zircon crystals from ancient sandstone and conglomerate outcroppings (conglomerate is a type of sedimentary rock that contains particles of a range of sizes, from small sand grains to large pebbles and sometimes even small boulders). They began this work before the great antiquity of the Acasta Gneiss was known; at the time, the oldest rocks that had been identified—which came from western Greenland—had ages near 3.8 billion years. The Australian sedimentary rocks, which outcrop in the western part of the country, had not been dated directly; however, nearby Precambrian gneiss had an age close to 3.6 billion years. By inference, it was thought the sedimentary rocks might be similarly old. If that were the case, any zircon grains they contained would be even older.

The painstaking work paid off. Most of the Australian zircons had ages of 3.5 to 3.75 billion years; but four grains gave dates between 4.1 and 4.2 billion years. Clearly this was older than any known existing fragment of the Earth's crust. Because zircons usually form in igneous rocks similar to granite, it was concluded—and it is still generally agreed today—that these crystals must have been eroded from granite or granitelike rocks. This was a significant finding. No traces of the parent rocks have been found—they have been destroyed by geological processes over the billions of years since. But the zircons, eroded from those granites eons ago and deposited in sandy sediments along an ancient shoreline, bring us the message that they once existed. They tell us that only a few hundred million years after the Earth formed, geological processes were creating crustal rocks on the Earth that were not radically different from those forming today.

The Australian National University researchers were led by Bill Compston, an early developer of ion microprobes specifically designed for uranium-lead dating of zircons. They published their findings in the journal *Nature* in 1983. For geologists, it was sensational news. Suddenly, a tiny flicker of light began to shine on that unknown, pre-3.8-billion-year period of Earth's history. As the Acasta gneisses would do later, the rocks in western Australia drew researchers from around the world. Many of them collected samples, took them back to their laboratories, and began the lengthy process of separating out zircons for dating. Hundreds have been analyzed, and most have ages less than 4 billion years. But a few—and it is a very few—are older. One, a single crystal only about a hundredth of an inch across, is the most ancient grain ever to have been found on Earth. It is 4.4 billion years old.

That one result has tremendous significance, not only for geologists, but for anyone interested in understanding the history of our planet, because it gives us a peek at the conditions prevailing during the Earth's earliest days. The very existence of that zircon crystal tells us that the Earth's surface was cool enough for igneous rocks to congeal and harden when the whole planet was only about 150 million years old. In all likelihood, those rocks were not very different from the granite some people use for the countertops in their kitchens. And even though the single ancient grain is small, microanalysis techniques have been used to measure its chemical properties in addition to its age. Without going into the details, one conclusion that can be drawn from the results is that there was liquid water on the Earth's surface when the ancient granite formed. At first, that may not seem very surprising. We are all familiar with an Earth on which two-thirds of the surface is covered in water; why shouldn't it have been like that 4.4 billion years ago? But the interval of the geological time scale before 3.8 billion years ago is referred to as the "Hadean" (translation: hell-like) because it is generally agreed that it must have been extremely hot. The Earth was still growing, sweeping up material in its orbit around the sun, and the frequent impacts, especially the large ones, released large amounts of heat energy. The pockmarked surface of our neighbor the moon still preserves evidence of that bombardment; most of its impact craters date to 3.9 billion years or earlier. On the Earth, geological processes have erased the physical evidence, but, because our planet has a much larger gravitational field, we would have attracted far more orbiting debris than the moon. For tens of millions of years, perhaps even for 100 million years or more, it would have been much too hot for water to exist on the surface; some geologists have speculated that there was even a long period when most of the Earth's surface was molten. The importance of the Australian zircons is that they provide a marker for the cooling of the early Earth—and even a guide to the actual temperature, because they indicate that where they formed, the surface had cooled below the boiling point of water.

Thus, although we (so far) have no actual rocks to analyze from the first half billion years of the Earth's history, a single crystal probed by modern analytical techniques has provided us with a remarkable amount of information about the environment of that distant time. Some scientists have claimed that they could read a planet's history in a grain of sand. There is some exaggeration in that statement, but if the grain was a zircon crystal, it might not be far off the mark.

CHAPTER SIX

Dating the Boundaries

William Smith is a very common name, and it is likely that there were many William Smiths born in eighteenth-century England. Most of them probably led fruitful lives, worthy of more than passing interest. But, as far as geologists are concerned, there is only one who matters: the William Smith who virtually single-handedly put together the first true geological map of Britain ever produced. The author Simon Winchester wrote a book about Smith and his work, calling it *The Map That Changed the World*. Smith's map was the culmination of years of observation and investigation, and it was—and still is—a masterpiece. But it is the concepts behind the map that concern us here, because they relate directly to deep time. James Hutton may have been the man who recognized the immensity of geological time, but William Smith was the person who brought a sense of order to its passing.

The last few chapters have examined how age determinations using radioactive isotopes have quantified the opposite ends of the Earth's time scale: its beginnings 4.5 billion years ago, established through uranium-lead dating; and the (geologically) short period that immediately precedes the present, encompassing much of human history, established through radiocarbon dating. But we also have a detailed time scale for everything in between. We know, for example, when fish appeared in the sea and when animals climbed out onto the land; we know that about 1 billion years ago, in what is now eastern North America, there was a great mountain range rivaling today's Himalayas; and we know that an extensive ice age gripped the Earth about 2.3 billion years ago. How have these things been worked out? How did we acquire a detailed time scale that orders important geological events throughout the whole of Earth's history?

To examine those questions fully, and to provide a background to present-day dating studies, it is useful once again to travel back a few hundred years into the past, to the time before radioactivity was discovered, when the very word *geology* was not yet in use. It was into this world that William Smith was born, in 1769, to a village blacksmith in England. James Hutton was then forty-three and had not yet formulated his ideas about the Earth's great age. George III was king, and was having problems with those pesky colonies in North America. As a child, William Smith was aware of none of this, but he was, by all accounts, a good student in school, an avid reader, and keenly interested in rocks, the local landscape, and—fossils.

Fossils, as we will see, played a central role in Smith's life and his creation of a geological map. When he came into the world, the conventional wisdom about these strange artifacts was in a state of flux. For centuries it had been thought—at least in England—that fossils were clever imitations of living creatures, made by an omnipotent Creator either to confuse or to impress ordinary humans, depending on your point of view. Or at least that was the idea espoused by most religious leaders. It was bolstered by the fact that in some places fossils resembling marine creatures could be found on high hills or mountains. How else could they get there, so far above the sea? But it is unlikely that Smith, with his inquisitive mind and objective nature, would have believed that explanation. However, even most of those who thought that fossils were the remains of once-living organisms assumed they originated in the biblical flood, and that counterparts of all the fossilized creatures must have been present on Noah's Ark. Undoubtedly, Smith would have been exposed to this theory at some point, and he might, for a while at least, have thought it was true.

We do not know a great deal about Smith's early life. We do know that his father died when he was quite young, and that Smith was sent off to live with an uncle. And we also know that, in spite of being quite bright in school, there was little or no chance he would go on to university—his financial situation and social position virtually guaranteed he could not. Smith left school when he was eleven. But, when he was just eighteen, he began a career that would, after many twists and turns, lead to his being honored by the English scientific elite as one of the preeminent geologists of his day. It all began when, quite by chance, he met a professional surveyor who happened to be working in the region where he lived. Smith already knew something about surveying from his reading, and struck up a conversation about the subject. That was all it took; the surveyor was impressed and, more or less on the spot, hired Smith as his assistant.

It was an excellent move for the surveyor. Smith was a quick learner and soon developed a reputation for being efficient and precise in his work. Before long, his employer was sending Smith off on his own to complete surveys, drill boreholes for coal, and devise plans for draining boggy farmland. During all this activity—which entailed much travel around the countryside—Smith developed an intimate knowledge of the topography and, just as important for his later work, the nature of the underlying rocks. He quickly came to realize that it was this underlying bedrock that actually shaped everything on the surface: it affected the soil, the vegetation, the drainage, and the topography.

If Smith consciously had one of those Eureka! moments, it must have come when his work took him to several coal mines in the southwest of England, near the town of Bath. As is often the case in such circumstances, his discovery was in some ways not a discovery at all, because the local miners already knew much of what he was to learn. But it was Smith who eventually put it all in context and saw the bigger picture. As he descended into the mines, he made notes about what he viewed:

initially just layers of sedimentary rock, lying almost horizontal, but all, he observed, tilting slightly in one direction. Then, farther down, there was an abrupt change; the layers suddenly looked very different. For one thing, they were no longer almost horizontal but dipped steeply downward. They were also more contorted, cracked, and dislocated, and sometimes they were folded over on themselves. They just looked *older.* It was as though some great force had affected these lower rocks, but not the ones lying on top. The boundary was sharp; it was an unconformity, just like the one Hutton had described at Jedburgh (see figure 2, on page 10). And, like Hutton, Smith realized that the unconformity separated rocks from different periods of geological time.

The coal seams occurred as a series of layers within the older-looking rocks that lay below the unconformity. Through long familiarity, the miners knew each seam intimately and instinctively in the same way you or I might immediately recognize the make of a passing car, or know the batting averages of the players on our favorite baseball team. They also knew—and Smith quickly learned—that the same sequence of layers was repeated in each mine pit. The folding of the deeper layers made it a bit complicated, but each identifiable seam of coal always had the same layers of sedimentary rock above and below it, recognizable by color or texture or—in some cases—by the fossils they contained. The whole complicated structure, with the rock strata piled up like a slightly crumpled stack of pancakes, was continuous between the pits. Connect the dots, and you could make a three-dimensional map of the coal seams, and all the other layers, too. And, just maybe, by extrapolating that three-dimensional picture over long distances, you could predict the types of rocks that would occur miles or even tens of miles away.

That first inkling Smith had that he could map out the geology by tracing individual layers of sedimentary rocks was—although he didn't know it—based on a principle that had been proposed more than a century earlier by a Danish-Italian scholar named Nicolaus Steno. It is a principle so simple that today it hardly seems worth stating, but in Steno's day it was a revelation, although it was still little appreciated in

Smith's time. Steno said that, because sediments are deposited from bodies of water, all sedimentary rock layers, regardless of their present appearance, must originally have been laid down horizontally. He also noted that younger sedimentary layers would always lie on top of older ones. Those simple statements are still basic rules of field geology. They can sometimes be difficult to apply, especially in places like the Alps, where the forces of mountain building have created great folds in the sedimentary layers, and in places have even flipped them over so that what is now "up" was once "down." But there are ways to sort out even those complications, and skilled geological mappers have no problem reconstructing the original sequences based on Steno's principles.

Smith soon had a golden opportunity to test his germ of an idea about the continuity of rock strata. Coal from the mines he was working in had to be taken to cities or ports for sale, and, in England toward the end of the eighteenth century, horse-pulled canal boats were the preferred mode of transport—they were the freight trains of their day. But there was no canal serving the mines Smith worked in, and so plans were laid to build one. It would be called the Somersetshire Coal Canal, and it would be relatively short, connecting the mines to other, already-existing canals to the east. Smith was appointed surveyor-in-chief for the project.

As it turned out, the route chosen for the canal was serendipitously near-perfect for revealing the details of the region's sedimentary geology because it cut neatly across the dipping sedimentary rock strata. In addition, it was decided that the canal should have two branches running parallel to one another, a mile or two apart, joining together at one end. That meant Smith could observe the sedimentary layers in two different places—in each branch of the canal—and work out their orientation in three dimensions.

Smith's ability to visualize the underground rock strata in three dimensions—he once described sedimentary rock layers as resembling a series of slices of bread and butter—was the key to his success at geological mapping. It is a skill that still characterizes the best field mappers, although now they can put their data into a computer and view them from any perspective, whereas Smith had only his sketches and his brain to work with. In much the same way that the miners could recognize individual coal seams, Smith was usually able to identify individual sedimentary rock layers in different locations from their thickness, color, grain size, and other physical characteristics. But, over long distances, the thickness of a layer might change, or its color might vary. That was a natural result of geographical variations in the conditions under which the sediments were originally deposited. However, the same fossil organisms could usually be found in a particular layer, regardless of its other properties. As Smith expanded his geological observations, fossils became ever more important for correlating rock layers from one region to another. He traveled widely, and at every fossil-bearing locality he visited he would stop and take samples for later study. Eventually he amassed a huge collection from throughout the country. Smith took great pride in his ability to place any particular fossil from his collection correctly in the overall sequence of British sedimentary rocks.

It is worth pondering Smith's facility with the fossil sequence for a moment. When he amazed his friends and colleagues with this knowledge, it was well before the publication of Darwin's theory of evolution. Smith didn't put all his trilobite fossils in one drawer and all the ammonites in another, as most collectors did. Instead, he arranged them in a *time sequence.* The oldest fossils were at one end of his collection, the youngest at the other. Empirically, and without understanding the details of how evolutionary changes came about, Smith recognized that there are small but nevertheless distinctive variations that distinguish fossils from different rock strata and therefore different time periods. And, in addition to following changes in individual fossil organisms, he also recognized that the overall assemblage preserved in a particular layer could be a correlation tool and time indicator. Species have come and gone continuously through geological time, and each time slice is characterized by a specific and often unique assemblage of fossil organisms in sedimentary rocks. This characteristic allowed Smith to place strata from far-flung localities in the correct relative time sequence.

Darwin later proposed that natural selection is the primary mechanism for evolution; Smith was using the results of that mechanism for his correlation and mapping.

Employing fossils in this way was a great leap forward for geology. It was also something of a surprise for geologists, used as they were to dealing with inanimate rocks, to realize suddenly that biology was important for their field. Years after Smith's work became widely known, Charles Lyell, writing about the science of geology in the first paragraph of his *Manual of Elementary Geology,* alluded to this surprise: "What is still more singular and unexpected," he wrote, "we soon become engaged in researches into the history of the animate creation, or of the various tribes of animals and plants which have, at different periods of the past, inhabited the globe."

Smith was a surveyor and engineer by trade, but his passion was observing and describing the rocks he found as he went about that work. He took on projects all over the country, traveling continuously, which allowed him to trace strata and collect samples from throughout Britain. In 1815, after years of work verging on obsession, Smith, then in his mid-forties, brought his extensive knowledge and observations together and published his geological map of Britain. It covered most of the country, depicting the location and nature of every major rock type and stratum. Smith himself hand-colored each rock unit on the map, and the result was magnificent. The map was huge, measuring more than eight feet high and six feet wide. It was also, incredibly enough, virtually a one-man accomplishment. Smith had had various assistants along the way, but no real collaborators. Single-handedly, he had mapped the entire country accurately and comprehensively. If you compare Smith's map with a twenty-first-century version, the similarity is uncanny.

Shortly after his map appeared, Smith published what was meant to be accompanying material, which he titled *Strata Identified by Organized Fossils.* Originally, he envisioned six separate volumes of this work, but in the end only four were completed. Each dealt with part of the rock sequence shown on his map, illustrating in great detail the characteristic

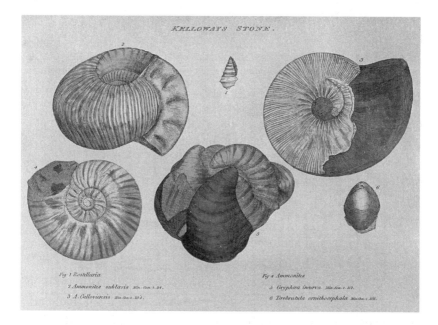

Figure 16. One of the illustrations from William Smith's *Strata Identified by Organized Fossils.* Although it may be difficult to discern, the page is titled "Kelloways Stone," which is one of the rock units on Smith's map. The three large coiled fossils are ammonites. From the website "William 'Strata' Smith on the Web" (www.unh.edu/esci/wmsmith.html). Reproduction courtesy of Professor Cecil Schneer, University of New Hampshire.

fossils of the different sedimentary strata. In a way, it was a kind of handbook. By matching fossils in a rock outcrop with those in Smith's *Strata Identified,* anyone could determine the outcrop's exact position in the overall sequence of British sedimentary rocks. And they could do it, as Smith himself put it, "without the necessity of deep reading, or the previous acquirement of difficult arts." The illustrations, like the map, were all beautifully colored by hand. An example is shown in figure 16, although, of necessity, in black and white. You can get a feeling for the full impact of the colored versions by viewing them on the website created by Professor Cecil Schneer of the University of New Hampshire, which also includes a reproduction of Smith's map.

Smith is sometimes credited with making the first geological map. That is not strictly true—many scientists and "natural historians" had traced sedimentary strata and mapped out the geology of various (usually local) regions of Europe long before Smith. In France, Georges Cuvier, who was born in the same year as Smith and became one of France's greatest biologists, published (together with his colleague Alexandre Brongniart) a geological map of the Paris Basin in 1808. But although it showed a sequence of sedimentary strata, and traced them across the basin, there is little evidence that Cuvier and Brongniart appreciated the dimension of time that was central to Smith's work. The Paris Basin map also covered only a fraction of the area that Smith's map did.

In spite of the importance of Smith's work—or perhaps because of it and out of professional jealousy—a small coterie of influential, educated, upper-class "geologists" in England conspired to cast aspersions on Smith's work while simultaneously plagiarizing it, causing him much personal and financial hardship. He was thrown into debtors' prison, and his London home and his belongings were seized for his debts. For many years after, he scraped together a living with odd surveying jobs around the country. But his genius could not be denied forever, and recognition, when it came, was great. In 1831, the Geological Society of London chose Smith as the first recipient of the Wollaston Medal, a prestigious award still given today. His work was praised effusively at the ceremony, and the president of the society referred to him as the "father of English geology." Later he was awarded an honorary degree by Trinity College, Dublin, and, more recently, a crater on Mars was named after him. In the words of William Berry, a professor at the University of California at Berkeley who has written about the making of the geological time scale, "William Smith provided geologists with a key by which the doors of past time might be unlocked."

Smith had toiled in relative isolation, and he worked only in Britain. But, at about the same time, and especially during the years immediately following publication of his map, other—mostly academic—geologists began to do similar work in Europe, the Americas, and Britain itself.

Unlike Smith, most of them were members of the intellectual elite, and they corresponded, read one another's papers, and traveled to observe the geology of other countries. Many of these workers conducted detailed studies of the sedimentary strata in their own regions, paying particular attention to the fossils these rocks contained. When their observations were compared across regions, and even from country to country, it quickly became apparent that Smith's principles for correlating rock units using fossils were universal. When Cuvier and Brongniart updated and republished their 1808 map of the Paris Basin in 1822, they adopted Smith's ideas, emphasized the succession of fossils in the sedimentary strata, and made direct comparisons to his work. Within a few decades, and still well before Darwin's celebrated *Origin of Species* appeared in 1859, geologists had constructed a time scale for the Earth's history using the principles Smith had outlined. Their time scale began when fossils first appeared abundantly in rocks—a time they labeled the Cambrian period, after the Roman name for Wales, where rocks of that age were first described in detail—and extended up to the present. They did not know when the Cambrian (or any other of the geological periods in the time scale) began or ended, nor did they know, as we do now, that most of the Earth's history actually occurred long before the Cambrian period. But, by systematically studying fossils of the different organisms that had inhabited the Earth in the past, they were able to construct a *relative* time scale, subdivided into units that were recognizable nearly everywhere. They knew that trilobites, small crustacean-like creatures, were crawling around on the sea floor before there were any fish, and they knew that both trilobites and fish existed long before there were any mammals or trees on the land. They knew that most of the living animals with which they were familiar—including humans—appeared only very recently in their relative time scale.

As a result of this work, the geological time scale was quite well established by the middle of the nineteenth century. In its general features, it was not radically different from that in use today, except, of course, there were no dates attached to it. Table 1 compares a modern version with one that appeared in an edition of Charles Lyell's best-selling *Manual of Elementary Geology,* published in 1852. The present-day time scale is more formalized than Lyell's, with various defined subdivisions (eras, periods, epochs), but the similarities to the earlier version are obvious. Today, geologists also recognize many additional subdivisions that are not shown here. A more complete geological time scale, including dates, appears as appendix A at the back of the book.

You can see from Table 1 that some of the subdivisions in Lyell's time scale were identified with a particular rock type, such as chalk, green-sand, or oolite (oolite is a sedimentary rock made up of small, cemented-together spherical particles). In Britain, these were the rock types that characterized these subdivisions, but, as you might imagine, the names were eventually changed to make them more universal. The major boundaries, and most of the lesser ones too, were defined by changes in the fossil assemblages that occur across them—and these are worldwide. The most dramatic changes define two major boundaries—that between the Paleozoic and Mesozoic eras, and that between the Meso-zoic and Cenozoic eras. The names of these geological eras reflect the magnitude of these changes: Paleozoic means "ancient life"; Mesozoic, "middle life"; Cenozoic, "recent life." The boundaries between these eras are now recognized as mass extinction events, geologically short periods when something—most likely some kind of environmental catastrophe—wiped out large fractions of life on Earth. The Paleozoic-Mesozoic boundary, generally known as the P-T boundary after the two geological periods that lie on either side of it (the Permian and Triassic; see Table 1), saw the largest mass extinction we know about. It has been estimated that up to 96 percent of all species then living became extinct, and some paleontologists have suggested that whatever caused this crisis came very close to eliminating *all* life on Earth. The boundary between the Mesozoic and Cenozoic, known as the K-T boundary (*K* for the German word for the Cretaceous, *Kreidezeit,* or "chalk time," and *T* for the Tertiary, the old name for the Cenozoic era), is slightly less dramatic in terms of the number of species that disappeared, but it has caught the

Table 1. *A Comparison of Lyell's (1852) Geological Time Scale with the Modern Version*

Lyell (1852)		Modern		
		Era	**Period**	**Epoch**
Post Pleistocene				Holocene
Pleistocene			Neogene	Pleistocene
Older Pliocene				Pliocene
Miocene	Tertiary,	Cenozoic		Miocene
	Supercretaceous,			Oligocene
	or Cainozoic		Paleogene	Eocene
Eocene				Paleocene
Chalk				
Greensand			Cretaceous	
Wealdon				
Upper Oolite	Secondary, or	Mesozoic		
Middle Oolite	Mesozoic		Jurassic	
Lower Oolite				
Lias				
Trias			Triassic	
Permian			Permian	
Coal			Carboniferous	
Devonian	Primary		Devonian	
Upper Silurian	Fossiliferous,		Silurian	
Lower Silurian	or Paleozoic	Paleozoic	Ordovician	
Cambrian &			Cambrian	
older fossiliferous strata				

imagination of the public because the extinctions included the dinosaurs and may have been caused, at least in part, by collision of a large asteroid with the Earth, and its associated effects.

After these mass extinctions, and also at the beginning of the Cambrian period (when another mass extinction may have occurred, although the evidence is not so clear), life proliferated again, albeit in different forms. Sedimentary rocks from these times are suddenly full of completely new types of fossils, things never seen in older rocks. Even a fairly cursory examination of fossilized organisms from either side of one of these boundaries produces convincing evidence that something quite drastic happened.

Initially the time scale had only three divisions: Primary, Secondary and Tertiary (Primary being the oldest). One of the first researchers to start subdividing it further was Charles Lyell. In the early 1830s he partitioned the Cenozoic era (still then known as the Tertiary) after studying fossils from sedimentary rocks in various parts of Europe. His approach was innovative. He separated the Tertiary into four parts based on a quantitative criterion: the percentage of fossils in each part that were recognizable as "living species of shells." He wrote, for example, that in rocks of the youngest subdivision, which he termed the Pleistocene, 90 percent of the fossils were known as living species, while in rocks from the next oldest subdivision, the Pliocene, only one-third to one-half had living equivalents. In the oldest rocks, right at the beginning of the Tertiary, there were very few fossils recognizable as present-day species.

The difficulty with Lyell's subdivisions was that their boundaries were vague. How could you find the exact boundary between Pleistocene rocks, in which 90 percent of the fossils were supposed to be similar to living species, and the Pliocene, in which only 30 to 50 percent fell into that category? Where would you place rocks in which three-quarters of the fossils matched present-day species? Although Lyell's scheme was followed for a while, more definitive criteria eventually emerged. Today, boundaries in the time scale are formally defined on the basis of the fossil record at a particular location in the world—the "type locality"—agreed on by an international body of experts. Usually it is a place where the sedimentary rock sequence is continuous, and where the exact location of the boundary is marked by a distinct biological event—for example, the appearance of one or more new fossil species. From that one location, what is sometimes referred to as a "web of correlation" can be spread across the globe, connecting one region to the next until it is possible to recognize exactly where the boundary occurs almost anywhere. And, although the whole approach has been brought up to date to encompass twenty-first-century knowledge and technology, the principles that William Smith devised to make his geological map still underlie the process.

The geological time scale is arguably geology's most important contribution to human knowledge, because it provides the framework for understanding the Earth's long history. But it was apparent from the beginning that there was a problem: rock units were assigned a place in the time scale on the basis of the fossils they contained, and there were abundant rocks that contained no fossils at all. The question was how to fit these rocks into the relative scale. In many places, they lay *beneath* rocks of the oldest part of the time scale, the Cambrian period, and they therefore had to be even older. Initially they were simply referred to as "Precambrian," and no attempt was made to subdivide them.

However, once dating with radioactive isotopes became possible, the full span of the Precambrian—accounting for roughly seven-eighths of the Earth's history—became apparent, and it seemed reasonable to subdivide it. The difficulty was how to do that. With no fossils (or at least very few—we'll come to that shortly) to mark out different times, there seemed to be little choice beyond arbitrary numerical divisions. But that was not a very appealing solution when each boundary in earlier parts of the time scale was defined in the rocks by a clearly observable event.

Worldwide field studies of Precambrian rocks convinced some geologists that there is a discernible change in chemical compositions and rock types near 2.5 billion years ago, and for that reason 2.5 billion years has become the major dividing line within the Precambrian—the boundary between the Archean ("ancient") and Proterozoic ("young life") parts of the Precambrian. Recently, in 2004, a new geological period, the Ediacaran, was formally added to the time scale. It immediately precedes the Cambrian period, and therefore lies within the Proterozoic. It is the first and only subdivision of the Precambrian to be based on fossils. Although they are rare and not very obvious, it turns out that, after all, there are some fossils in the Precambrian.

To early geologists, the start of the Cambrian period, which has now been precisely dated at 542 million years ago, was the time when life appeared on Earth. That, at least, is how it seemed, because as far as these early researchers could tell, Cambrian rocks were the oldest in which fossils occurred. But, as time went by, this conclusion began to be questioned, because the earliest fossils were actually quite complex. The organisms were small, and they all lived in the sea, but they had shells and skeletons and many advanced biological attributes. How could they just appear, so fully formed and in such great variety?

In fact, there was life in the seas long before the Cambrian period, but the organisms were soft bodied, lacking shells or other hard structures (see figure 17). As a result, they were only preserved under very specific and quite rare sedimentary conditions, and don't often occur as fossils. Some are known only from the tracks and traces they left in ocean-bottom mud. One of a small number of places in the world where these organisms are well preserved is in the sedimentary rocks at Ediacara, Australia. They seem from their fossils to have been strange creatures, some wormlike, others resembling fronds attached to the sea floor on small discs. Some have been described as looking like miniature air mattresses. Collectively, they are referred to as the Ediacaran biota, and the new geological period was named after them. When it was formally adopted in 2004, the Ediacaran period was the first such addition to the geological time scale since 1891—obviously a big deal. In 1891, of course, there were no dates for time scale boundaries. In our current understanding, the Ediacaran period lasted from approximately 630 million years ago until the beginning of the Cambrian period, 542 million years ago.

Figure 17. The creatures of the Ediacaran period, the oldest complex organisms known on Earth, were soft-bodied animals, and their fossils are often not as distinct as those of later animals with shells and bones. This example, known as Charnia, is characteristic of the Ediacaran fauna and is found in rocks ranging in age from about 575 to 553 million years. It has been described as looking like the frond of a fern, and in life (like other Ediacaran organisms) it was "quilted." The partial frond shown in the photograph is about 1.25 inches wide. This fossil is a University of California Museum of Paleontology specimen from the Winter Coast of the White Sea, Russia, collected and photographed by Jere H. Lipps.

Figure 18. Researcher Chris MacIsaac, garbed in lint-free lab clothing and working in an enclosure bathed in filtered, laminar-flow air, prepares samples for isotope analysis in a clean lab at the Scripps Institution of Oceanography, University of California at San Diego. Photo courtesy of Pat Castillo.

That brings us to the question, How exactly *are* the boundaries of the time scale dated? Ages for the Cambrian-Ediacaran and younger boundaries are believed to be accurate to within about a million years, or less. That is quite remarkable—it is an accuracy of 0.5 percent, or better. Such precise age measurements have been possible only fairly recently. In part, they are the result of advances in scientific instrumentation. In part, they stem from the recognition and control of laboratory contamination, as Clair Patterson did with lead. You can't just stroll into a modern dating laboratory—the chemical processing of samples and much of the other work takes place in "clean labs," with researchers garbed head to toe in lint-free garments and all incoming air filtered to remove even very small dust particles (see figure 18). But, to a large degree, the accuracy of boundary dating also comes down to finding the right kind of samples in the right place.

One of the difficulties in determining the ages of geological boundaries is that they are defined by fossils in sedimentary rocks, and sedimentary rocks are notoriously difficult to date. It is easy enough to understand why—they are typically made up of particles eroded from the continents and transported to the sea, which means that each grain in these rocks has a history, quite possibly a long history, prior to its deposition. The ancient zircons discussed in chapter 5, from Precambrian sedimentary rocks in Australia, date the crystallization of their precursor granite, not the formation of sediments that incorporated them. There are some sedimentary rocks that are not made up of older particles—a good example is limestone, which can form either by direct precipitation from seawater, or as an accumulation of small shells. But, when it forms, limestone doesn't incorporate any of the radioactive isotopes necessary for age determination. This is particularly unfortunate because limestone is one of the most widespread sedimentary rock types and frequently contains the very fossils used to construct the relative time scale.

In spite of this apparently gloomy outlook, nature has provided several ways to circumvent what might at first seem to be an insurmountable problem. One of these is to use indirect methods of dating— for example, by measuring a property of the rocks that varies in a known way over time. A very successful indirect approach is based on changes in the Earth's magnetic field, which are faithfully recorded in some sedimentary rocks when they form. Because the timing of these changes is known independently, the rocks can be dated simply by measuring their magnetic properties.

However, perhaps surprisingly, the single most useful approach for dating sedimentary rocks turns out to be related to volcanic eruptions. Active volcanoes are numerous and geographically widespread, and somewhere on Earth there is a volcanic eruption going on nearly all the time. As far as we know, this has been true throughout geological history. One of the consequences of plate tectonics is that the most explosive volcanoes, like those of Indonesia or the Philippines, occur around the margins of the oceans. When these volcanoes erupt, the effects are not only locally devastating but are often evident thousands of miles away. Fine particles of volcanic ash are thrown high into the atmosphere and may be carried halfway around the world by winds. When the particles of ash settle, and especially if they settle in the ocean, they form a discrete, identifiable layer. Best of all, a volcanic ash layer is a time marker, because its deposition over a wide area is, in a geological sense, instantaneous.

Volcanic "ash" is not really ash in the conventional sense, although at first glance it looks just like the stuff you sweep from your fireplace after burning a few logs. In reality, though, it is composed of small grains of the same material that makes up the volcano from which it came—that is, the same minerals present in the volcanic lava also occur in the ash. Some of these minerals—like zircon, which is relatively common—are suitable for age determination. And, in many volcanically active regions, large eruptions occur every few thousand years, producing an abundance of potentially datable layers. Such settings have provided some of the most accurate ages available for boundaries of the geological time scale. Even if there is no ash layer right at the boundary, it can often be bracketed precisely with dates on ash deposited before and after the boundary layer itself.

Several different radiometric dating methods have been used to date minerals from volcanic ash layers. The most important are listed in Table 2, which shows that each involves a different radioactive parent and

Table 2. *Principle Dating Methods for the Distant Past*

Method	Radioactive Parent Isotope	Stable Daughter Isotope	Half-Life
Potassium-Argon	potassium-40	argon-40	1.25 billion years
Rubidium-Strontium	rubidium-87	strontium-87	48.8 billion years
Uranium-Lead	uranium-238	lead-206	4.47 billion years
	uranium-235	lead-207	704 million years

stable daughter isotope. Together, the elements involved in these decay schemes span a range of chemical characteristics and behaviors, features that determine which technique is best for a specific mineral or rock type. Zircon crystals are ideal for uranium-lead dating because of their high uranium content, but they can't be dated using the rubidium-strontium method because they don't incorporate much rubidium. The half-lives of the parent isotopes in these techniques range widely too, from less than 1 billion years for uranium-235 to almost 49 billion years for rubidium-87. Obviously, these half-lives are all far greater than the carbon-14 half-life of 5,730 years. That means they can be used to measure ages throughout nearly the whole of the Earth's history, with the exception of the youngest part. Trying to determine the age of a 1-million-year-old sample using rubidium-strontium dating would be virtually impossible because over that (geologically) short period of time, too few rubidium atoms would have decayed. On the other hand, with care, it is possible to date 1-million-year-old and even much younger samples using the potassium-argon method, which has a significantly shorter half-life (the method also has some additional characteristics that make it especially useful for dating young samples).

As already mentioned, uranium-lead dating has a special feature: there are two different uranium isotopes that decay to produce two different stable isotopes of lead. This is the property that allowed Patterson to measure the age of the Earth simply by measuring the ratio of the two lead isotopes, and it also provides a kind of built-in cross-check for dating. If ages calculated from each of the two decay paths agree, it is likely that the date is correct. This feature makes uranium-lead dating the most widely used technique for high-precision age measurement.

But that does not mean that other methods play only a subsidiary role; each of the dating methods listed in Table 2 is in common use. In most cases, samples are analyzed using only one of these techniques. However, if the reliability of an age determination is in question, it can be tested by repeating the measurement using a different technique. In general, processes that might affect an analysis—for example, natural chemical alteration of a sample—affect the chemical elements involved in the different dating methods differently, leading to disagreement between the dates. Thus agreement between techniques provides a high degree of confidence in the results.

The utility of using multiple techniques was shown dramatically in work by H. Baadsgaard, a professor at the University of Alberta, Canada, and his colleagues in 1993. These researchers used all three methods listed in Table 2 to date samples from a volcanic ash layer in sedimentary rocks from western Canada. By separating different minerals from the ash layer, Baadsgaard was able to find material suitable for each of the dating techniques—zircon crystals for uranium-lead dating, potassium-rich mica for potassium-argon dating, and feldspar crystals for the rubidium-strontium method. When the researchers compared their results, they found no discernible difference among the dates. Three completely independent techniques gave ages of 72.5, 72.5, and 72.4 million years, in each case with a measurement uncertainty of just a few tenths of a million years.

Baadsgaard and his colleagues chose this particular ash layer for dating because it lies adjacent to a boundary between two subdivisions of the Cretaceous period. The boundary is defined by the appearance of a new fossil species in the geological record, and the researchers' result—they averaged the data and assigned an age of 72.5 ± 0.4 million years—significantly improved the existing understanding of exactly when that biological event occurred.

Performing multiple age analyses is a time-consuming and expensive proposition. Often it involves several different laboratories because geochronologists tend to specialize in only one technique. For these reasons, it is a fairly rare occurrence. But the example just described does illustrate an important aspect of the geological time scale: it is continually being refined. In a sense, it is a work in progress. As methods of analysis improve and as rock exposures in the field are studied more intensively and in more places around the world, the ages of boundaries are refined and updated. Most of these changes are small. However, an important trend resulting from the advances in analytical techniques is that the uncertainties associated with radiometric dates are gradually decreasing, making it possible to examine events in the Earth's history with increasingly higher resolution. This can be crucial for understanding the causes of the biological events that mark boundaries in the time scale.

A case in point is the Permian-Triassic boundary, which, as already pointed out, marks the most dramatic biological extinction that has ever occurred on Earth. The type locality for the boundary is at Meishan, China, near the city of Nanjing. There, in a series of now-abandoned rock quarries, the dramatic change in fossil assemblages is clearly delineated in the sedimentary strata. Furthermore, sandwiched within the rock layers are abundant beds of volcanic ash.

The ash layers at Meishan have been dated by all three methods in Table 2: potassium-argon, rubidium-strontium, and uranium-lead. In general, the results have been quite consistent, falling close to 250 million years. But, until 1998, when Sam Bowring and his colleagues from MIT reexamined the problem and published new uranium-lead dates, the smallest uncertainty on an individual age determination was ± 1.5 million years. This was for a potassium-argon date made on mineral grains from an ash layer that lies just below the P-T boundary at Meishan. It was, by any standards, a very precise result, with an uncertainty of less than 1 percent. Even so, it was not precise enough to distinguish between possible causes for the extinctions. Were they due to a sudden, catastrophic event like the asteroid impact at the K-T boundary, or were they the result of longer-term environmental changes that might occur over a million years or more?

In an attempt to improve this situation, Bowring and his colleagues carried out almost two hundred uranium-lead analyses of zircon crystals from the ash layers at Meishan and two other localities in China. By bracketing the boundary with dates on closely spaced ash layers, they were able to assign a precise age of 251.4 million years, with an uncertainty of only ± 0.3 million (300,000) years. That was a truly remarkable achievement; it put new constraints on the duration of the P-T extinctions. The most recent (2004) official version of the geological time scale, which integrates the best age information from a variety of sources, lists the P-T boundary date as 251.0 ± 0.4 million years.

However—there always seems to be a *however,* and it emphasizes what I said a few pages back about the time scale's being a work in progress—there are indications that even this date may be slightly in error. A new technique has been developed for treating zircon crystals before they are analyzed, one that seems to solve a problem that sometimes affects uranium-lead dates: "leakage" of small amounts of lead from the crystals over time. Even tiny amounts of lead loss can affect the very precise dating that is possible with modern analytical methods, making the ages too young. Roland Mundil, of the Berkeley Geochronology Center in California, found that heating zircons to high temperatures and then etching them with strong acids before analysis is an effective way to remove portions of the crystals that have been affected by lead leakage. Exactly why this treatment works the way it does is not entirely clear, but Mundil finds that it produces very consistent results, even for zircons known to be affected by lead loss. In 2004, he used the new method for samples from the Chinese P-T boundary site, and got an age of 252.6 million years, very slightly older than the Bowring result and the "official" date on the 2004 time scale. The uncertainty assigned to the new date is only plus or minus 200,000 years, and Mundil believes that the main pulse of severe extinction occurred entirely within this short time interval. This clearly narrows down the range of plausible extinction mechanisms to ones that act very rapidly.

The precise age for the Meishan samples dates the biological events recorded in the rocks at a particular locality. But the world is very large, and there is always a question of synchronicity. Did the same extinction (or appearance of a new species) take place at the same time everywhere? For mass extinctions like those at the P-T or K-T boundaries, worldwide synchronicity is probably a valid assumption. But, for boundaries that mark the finer subdivisions of the time scale, the picture is not so clear. One can imagine some small marine organism becoming extinct along the California coast because of changing water temperatures but surviving for a period of time, in much reduced numbers, in a similar habitat in Chile. Such things have been documented in the past, and geologists refer to them as being time-transgressive: events that occur at different times in different places. If the overall time periods are short, high time resolution is required to determine whether an event is truly time-transgressive. And if high time resolution cannot be achieved by direct dating, indirect methods may have to be used.

Anything that occurs geologically instantaneously and is recorded in sedimentary rocks over a wide area is ideal for examining the question of time transgressiveness. Volcanic ash layers fall into this category, although only those from the very largest eruptions are traceable over great distances, and none provide a truly global time marker. There is, however, an important characteristic of sediments that *is* imprinted simultaneously worldwide: their magnetic properties.

In the early 1950s it was discovered that the Earth's magnetic field—which approximates the field that would exist if there were a giant bar magnet embedded in our planet—periodically reverses polarity. Were that to happen today, adventurers would be in trouble, because their compass needles would point south instead of north. Geomagnetic reversals occur fairly quickly—over periods of a thousand or, at most, a few thousand years—and are usually preceded by a gradual weakening of the magnetic field. Some scientists predict that we are currently heading toward a geomagnetic reversal, because the strength of the Earth's magnetic field has decreased substantially over the past few centuries. The last reversal occurred 780,000 years ago (the timing determined by potassium-argon dating), and, although the time between reversals varies widely, on average there have been four to five reversals every million years over the past 10 million years. So, in a purely statistical sense, we are due for one.

Some of the iron-containing minerals found in rocks are magnetic, and, when igneous rocks crystallize from cooling magma, these minerals align their own magnetic fields with that of the Earth, just as though they contained their own little bar magnets. The effect is to "freeze in" a record of the orientation (and also the strength) of the Earth's magnetic field at the time the rock formed. The discovery of magnetic field reversals came when geophysicists studying the magnetic properties of lava flows noticed that some flows seemed to be magnetized in the opposite direction to that of the Earth's current magnetic field. At first they thought this was an artifact of some kind. But when they found exactly the same pattern of "normal" and reversely magnetized lava flows in different localities, it became obvious the reversals were a widespread natural phenomenon. The only possible cause, they realized, was that the Earth's magnetic field had periodically reversed in the past, and that the direction of the Earth's magnetic field had been faithfully recorded by each lava flow as it formed. This discovery initiated a new discipline within the earth sciences: paleomagnetism, the study of the Earth's magnetic field in the past.

The early work in paleomagnetism was carried out entirely on igneous rocks, which, as we have seen, can be readily dated by radiometric methods. By measuring rocks of different ages in different places, researchers gradually built up a detailed historical record of the Earth's magnetic reversals. As with the fossil-based time scale, scientists paid particular attention to dating boundaries—although in

Figure 19. The geomagnetic time scale back to 25 million years. "Normal" periods, when the Earth's magnetic field was the same as today, are shown in black; periods in which the Earth's magnetic fields were reversed are shown in white. Some intervals are so short that they do not show up at this compressed scale. Based on data from S. C. Cande and D. V. Kent, *Journal of Geophysical Research* 100 (1995): 6093–95.

this case the boundaries were different ones, marking the exact times of the magnetic reversals. And, unlike the fossil-based boundaries, in this case there was no possibility that the magnetic reversal boundaries were time-transgressive—they happened everywhere at the same time. The pattern of magnetic reversals—the "geomagnetic time scale"—is now well known back into the Jurassic period, more than 150 million years ago, and is a standard illustration in geology textbooks. With its black-and-white bands, it could pass for the barcode on a can of tomatoes (see figure 19).

Not long after magnetic reversals were discovered, scientists measuring the magnetic properties of deep-sea sediment cores found that, just as in the case of lava flows on land, there are intervals of both normal and reversed magnetism in sediments. It was soon realized that tiny grains of magnetic minerals, eroded out of rocks on land and washed into the sea, line up with the Earth's magnetic field as they slowly settle to the ocean floor. As a result, the magnetic reversal barcode gets imprinted on the accumulating layers of sediment. But, unlike lava flows, which are erupted sporadically, ocean sediments are deposited continuously and therefore retain an uninterrupted record. With the reversal boundaries well dated by a combination of studies of igneous rocks on land and volcanic ash layers in sediments, the age of any particular level in a core can now be determined quite accurately just by matching up the reversal patterns.

The time scale provided by magnetic reversals has been a great boon to those researchers intent on making finer and finer subdivisions of the most recent part of the geological time scale, the Cenozoic era. In the 1960s, at roughly the same time as NASA was preparing to land men on the moon, another ambitious project was launched in the United States. It was called the Deep Sea Drilling Project, and its aim was to use oil industry technology to drill deep into the sea floor in all the world's oceans. Until then, sampling of the ocean bottom had relied on dredges dragged behind ships, or on coring devices that were simply dropped through the water to penetrate sediments under the force of gravity. Neither did much more than scratch the surface of the ocean floor. The Deep Sea Drilling Project, in contrast, has retrieved cores that span millions or even tens of millions of years of geological time. In its several incarnations, it has brought back literally hundreds of miles of sediment cores that have revolutionized our understanding of the history of the oceans. The project has been so successful that it continues to this day, expanded to include wide international participation and now known as the Integrated Ocean Drilling Program.

A crucial task that faced researchers working on the drill cores was accurate dating, because only with reliable time information would it be possible to work out the history recorded in the sediments. For that reason, the magnetic properties of the cores are measured routinely, and the magnetic time scale applied. But it has also been possible to achieve even finer time resolution. In many of the cores, the sediments are composed nearly entirely of the microscopic shells of plankton, the tiny plants and animals that live their short lives floating in the ocean and that, when they die, create a continuous rain of biological debris onto the sea floor. Microscopic examination of literally hundreds of individual shells of these organisms—which paleontologists refer to as microfossils—from each small increment of the cores has provided unprecedented insight into their evolutionary changes. Systematic examination of a particular species of microfossil, for example, often reveals a minor change in the morphology of the organism's shell that might first appear in only one or two individuals of several hundred examined. Moving upward in the core, centimeter by centimeter, the proportion of shells with the new feature increases at the expense of the old morphology, until it reaches 100 percent. A new species is now firmly in place, the old one extinct. The period of overlap between old and new is generally very short in geological terms. By using the magnetic reversal time scale, supplemented where possible by precisely dated ash layers, scientists have been able to date many of these species changes, and to build up a new, densely subdivided and entirely fossil-based time scale. The fossil-based scale has been painstakingly assembled by identifying and assigning ages to small evolutionary changes in very large numbers of individuals from continuous sediment cores at multiple locations in the ocean basins, and it is now an invaluable resource for deep-sea sediment studies. If you give a specialist a speck of mud from the ocean, he or she will smear it on a glass slide, peer at it through a microscope, and quite quickly tell you how old it is. The microfossils thus provide yet another indirect dating method, one with very high time resolution. Often the fossil-based dates have even smaller uncertainties than would result from a single radioactivity-based date on the same sample (if it were even possible to carry out such an analysis), simply because microfossil evolution proceeds so rapidly. Nevertheless, all ages in the time scale, even for the very fine, fossil-based subdivisions of the Cenozoic, are ultimately tied to radiometric dating. Without radioactivity, neither the magnetic reversal patterns nor the evolutionary changes in microfossils could provide anything but a relative time scale.

CHAPTER SEVEN

Clocking Evolution

The end of the Cretaceous period came, in a twist on T. S. Eliot's famous line, not with a whimper, but with a bang. A very big bang that was felt around the globe, equivalent, it has been estimated, to the explosion of at least 100,000 billion tons of TNT, or about 8 billion times the energy released by the atomic bomb that devastated Hiroshima at the end of the Second World War. You would not want to have been around at the end of the Cretaceous.

The environmental crisis that ended the Cretaceous period and closed the Mesozoic era was of global extent, and, as we have seen, life on Earth changed abruptly. It was a time of mass extinctions—and also the beginning of a period of rapid biological evolution that saw the appearance of many new life forms. Dinosaurs disappeared, and, in a geologically short time, mammals became the dominant large animals on land (and in the sea). The detailed geochronology carried out at the K-T and other boundaries, and throughout other parts of the geological record as well, have profoundly impacted ideas about how evolution proceeds. Ever more precise dating studies have shown that the mass extinctions occurred over time spans of hundreds of thousands of years or less, not many million of years, as had been believed earlier. That evidence in turn has affected thinking about the importance of catastrophic events in changing the course of evolution. But chronological studies have also shown that, at other times in the past, evolution proceeded in fits and starts, not (as it is sometimes depicted) as a uniform, continuous process leading from ancient bacteria to *Homo sapiens*. Indeed, many paleontologists, prominent among them Stephen J. Gould, have challenged that conventional view of evolution, suggesting instead that natural selection, while very important, is only one of many factors that must be taken into account. Increasing complexity, they argue, is by no means a predictable outcome of evolution: our concepts of the process have been heavily skewed by our anthropocentric viewpoint. Precise dating of the fossil record is not the only basis for these conclusions, but it is an important part of the factual evidence that shapes them.

The catastrophic events that occurred at the K-T boundary and that were at least partly responsible for the extinctions were unknown until the 1980s. That was when Nobel Prize–winning physicist Luis Alvarez, together with his son Walter (a geologist) and several other colleagues, uncovered evidence that a very large comet or asteroid had struck the Earth precisely at the boundary. Like several other discoveries described in this book, theirs was unexpected.

At many places in the world, the K-T boundary is marked by a thin layer of clay that interrupts a long, continuous sequence of layer upon layer of limestone. The nature of the boundary had always been a bit of a puzzle. Limestone on either side of the clay layer contains fossil shells of marine organisms. The species are different, but the presence of similar types of fossils indicates that the environmental conditions were about the same both before and after the extinctions. Why then did the band of clay at the boundary contain virtually no fossils? Was it because the extinctions had wiped out nearly all life in the oceans, so the only thing being deposited on the sea floor was grains of clay washed in from the continents? And, if so, how long did this interval of a "dead ocean" last until the ocean plankton had reestablished themselves? It was this last question that particularly interested Alvarez and his colleagues. They thought that chemical analyses of the clay might provide a clue, because some chemical elements are known to accumulate more or less steadily in ocean sediments. If these elements had different concentrations in the clay layer compared with the layers above and below, it would be an indication that the clay collected at a different rate—and they would be able to calculate what that different rate was, which in turn would allow them to estimate the time span represented by the clay.

The Alvarez group first examined the K-T boundary at a well-studied locality in Italy. They analyzed both the boundary clay and samples from above and below it for twenty-eight different elements. Twenty-seven of them showed nothing unusual—their concentrations were roughly the same throughout the entire rock sequence. But there was one notable exception: the rare metal iridium was highly enriched in the clay compared with the limestone on either side. Thinking this might be a local anomaly, they measured another site in Italy and found the same thing. They then tested samples from Denmark and New Zealand, and again recorded high iridium contents in the clay layer at the boundary. Samples from opposite sides of the planet showed the same feature; it appeared to be a global phenomenon.

Most rocks from the Earth's crust are very poor in iridium, but it is well known that meteorites have concentrations of this element that are thousands of times greater. Alvarez came to the reasonable conclusion that the enrichments at the K-T boundary must be due to the collision of an asteroid, or perhaps a comet, with the Earth. A very large asteroid, six to seven miles in diameter, would be required to supply iridium at the levels they had measured if the layer was truly worldwide. An object of that size would have vaporized on impact, and all the iridium—and everything else it contained—would have been thrown into the atmosphere and spread around the globe, to drift down slowly over months and years and eventually be preserved in the K-T boundary clay.

The impact theory generated immediate controversy. Paleontologists who had been studying the end-of-Cretaceous extinctions for years claimed that even such a large impact couldn't possibly account for the worldwide extinctions, and, anyway, they said, their work showed that some species died out gradually, not instantaneously at the boundary. And there was the fact that in 1981 there were no known impact craters (of which there are about 120 on Earth) of the right size and age. Alvarez countered that argument: because two-thirds of the Earth is covered by oceans, it was quite possible—even likely—that the impact had occurred at sea and that the crater lay somewhere on the sea floor, unrecognized. An impactor that was six or seven miles in diameter would have ploughed through a few miles of seawater with ease.

Several decades on from the Alvarez discovery, most geologists are convinced that he was right, although not all agree that the impact was the sole cause of the extinctions. The smoking gun—the crater—has been discovered, buried under younger sedimentary rocks of the Yucatan Peninsula in Mexico. And, pertinent to our story, the timing of the impact has been precisely determined using potassium-argon dating.

It was realized as early as 1906 that potassium is radioactive, but it was not until almost thirty years later, in 1935, that the isotope responsible for its radioactivity (potassium-40) was detected by Alfred Nier, the University of Minnesota physicist mentioned in chapter 5 in connection with uranium-lead dating. Although other isotopes of potassium had been discovered much earlier, and theorists had already predicted that potassium-40 should exist, it makes up only 0.1 percent of natural potassium, and it took Nier's improved, sensitive mass spectrometers to detect it. However, Nier's instruments couldn't measure potassium-40's half-life, and nobody knew with certainty what its daughter product was. (The same theorists who had predicted potassium-40's occurrence had also predicted that it would decay to an isotope of calcium, calcium-40. However, this hypothesis had not been verified experimentally.)

Two years after Nier's discovery, the German physicist Carl Von Weizsäcker made an inspired suggestion. He noted that the inert gas argon is abundant in the atmosphere, and he inferred that it must be the daughter product of potassium-40 decay, which had accumulated there over geological time. He urged mass spectrometrists to analyze old, potassium-rich mineral samples to test his hypothesis. If he was right, they should find high levels of argon-40. But more years passed—it was, after all, the period of time that included the Second World War—until finally, in 1948, the experiment was done. Once again, it was Alfred Nier who made the measurements, along with his colleague Lyman Aldrich. They analyzed four different potassium-rich minerals, and in all of them they found excess amounts of argon-40. This result verified Von Weizsäcker's prediction, and it also laid the foundations for potassium-argon dating in much the same way that Libby and Anderson's detection of carbon-14 in Baltimore sewage gas had prepared the way for radiocarbon dating. But there was still much hard work ahead before this new method became routine; among other things, it was necessary to determine potassium-40's half-life accurately, and to develop procedures for making quantitative measurements of the potassium and argon contents of geological samples.

Aldrich and Nier's 1948 work also showed that the decay of potassium-40 is not straightforward. A few radioactive isotopes decay to more than one daughter isotope—and potassium-40 is one of them. The behavior is known as "branching" decay, and, in the case of potassium-40, the two daughter products are isotopes of argon and calcium, argon-40 and calcium-40. Because the fraction of decays that produces each daughter isotope is always the same and has been precisely measured experimentally, branching decay does not pose any problems for potassium-argon dating—it can be taken into account in the equations used to calculate the age of samples.

You might wonder why the calcium branch of the decay is not used, too; the answer is that calcium is so abundant in rocks and minerals that the amount added by radioactive decay of potassium-40 is miniscule by comparison, and hard to distinguish. In contrast, argon is inert and not incorporated into minerals when they form, so it can be safely assumed that virtually all the argon in a sample has been produced by radioactive decay. This makes the potassium-argon dating method very sensitive.

By the mid-1950s, researchers at a number of institutions—notably the University of California at Berkeley, the University of Chicago, and the Carnegie Institution of Washington—were working to realize the promise of potassium-argon dating. It was, at that stage, not nearly as well developed as uranium-lead dating, but constant improvements were being made. Initially, the scientists focused mostly on very old (Precambrian) rocks, partly because this part of the geological time scale, with no fossils to guide geologists, was ripe for age determinations, and partly because the older rocks contained more argon and were therefore easier to analyze. As the work progressed, several important characteristics of the potassium-argon method became clear. One was that the technique is especially useful for volcanic rocks, because when they erupt at the surface, any argon already present in the molten lava quickly escapes into the atmosphere, and the newly formed crystals of the volcanic rock start life with a clean slate, their radioactive clocks set to zero. Furthermore, because potassium is one of the most abundant chemical elements in the Earth's crust, most volcanic rocks contain minerals rich in potassium, which are ideal for dating. In addition, the relatively short (in geological terms) half-life of potassium-40 (see table 2 on page 150) meant that the method could be used for quite young rocks in addition to very old ones. "Quite young" here initially meant a few million years, but this limit quickly came down with improvements in technology and sample handling techniques until, by the 1990s, it was within the range of carbon-14 dating, tens of thousands of years. Measuring ages of such young material is by no means routine, but it is feasible in some circumstances. In a very few instances, researchers have even reported accurate dates of less than 10,000 years with potassium-argon dating. The applicability of this method to geologically young samples has led to its widespread use in archaeology and studies of human evolution and, together with its usefulness for dating volcanic rocks, has also made potassium-argon dating the primary technique for calibrating the magnetic reversal time scale.

But, to return to the K-T boundary (the reason for the brief detour into the details of potassium-argon dating will, I hope, become clear below): The Alvarez impact theory prompted geologists to reexamine the kinds of sediments preserved at the boundary. At localities close to the Yucatan Peninsula, not far from the crater, they found abundant, obvious fragments of ejecta thrown out when the asteroid struck. But there was also similar material much farther afield. Crystals that once resided in the solid crust of Mexico were found in boundary layer sediments as far away as Europe, Colorado, and the middle of the Pacific Ocean. Most commonly, these were small grains of quartz, crushed and deformed by the blast of the impact. But other minerals were found, too, including zircon crystals.

One site where K-T boundary layer sediments contain zircons is in Colorado. Here the zircon crystals are small, and, like the quartz crystals, they show shock features that link them to the impact. Another feature that links them is their age. Uranium-lead dating of these grains gives a date of 545 million years, much older than the age of the K-T boundary itself (which, as explained below, is 65.5 million years). But the age of the Colorado zircons matches exactly the age of zircons separated from the igneous rocks at the Yucatan impact site. The resilient zircons, apparently, survived the shock and heating of the impact, and their trip to Colorado, unscathed, their internal clocks intact. The age of the Colorado zircons didn't provide a date for the K-T boundary, but it did unequivocally link the zircon grains, and, by association, other material in the Colorado boundary layer, directly to the crater in Yucatan.

Other materials, however, did provide an accurate age for the impact. Throughout the Caribbean region, the ejecta layer is thick and easily identified. It contains abundant shocked quartz and other crystals, but it also contains tiny pieces of natural glass, usually spherical or teardrop shaped, and typically no more than one or two millimeters across. The Caribbean spherules are very similar to glassy droplets formed when molten lava is sprayed violently into the atmosphere by an erupting volcano, but they are not volcanic in origin. Their chemical compositions show that they are frozen droplets of the molten material thrown up when the asteroid crashed into and melted parts of the Earth's crust. The spherules contain large amounts of potassium, and it was quickly realized that these direct products of the impact would be ideal for potassium-argon dating. Several different laboratories have now analyzed them, and within the measurement uncertainties, the results are all the same. The glass spheres, and therefore the K-T boundary, have an age of precisely 65.5 million years.

There is a caveat that must be added to the previous sentence. If you were to look up the original scientific papers, you would find that the ages reported actually differed slightly—one study gave a date of 64.5 million years, another 65.0 million years, and both of these results are marginally lower than the currently accepted age for the boundary, 65.5 million years. The differences are not very large, yet they are outside the range that would be expected from the experimental uncertainties. For geologists seeking the most accurate dates available, this is an uncomfortable situation. In fact, for anyone not familiar with the intricacies of the dating methods, such untidy details are confusing, and they can cause doubts about the fundamental veracity of the techniques.

There is, however, a perfectly reasonable explanation. It has to do with the nitty-gritty of the dating method, and the fact that geochronologists— like all scientists—are constantly striving to improve their techniques and procedures. I will try to summarize without straying too far into jargon.

Except in the special case of radiocarbon dating, the dating methods discussed in this book generally require measurement of both parent and daughter isotope abundances in a sample to calculate its age. For the potassium-argon technique, this means measuring both potassium-40 and argon-40. In the conventional approach, these measurements had to be taken using two different instruments, because potassium is a solid and argon a gas. Each sample for dating was therefore split into two parts: potassium was measured using one instrument on one part of the sample, and argon gas was extracted from the other part of the sample by heating, then measured using a second instrument. But this procedure has the potential for introducing considerable error. To determine the concentration of any element or isotope requires that the sample be accurately weighed, and weighing out two portions rather than one doubles the uncertainty associated with that measurement. There is also the very real possibility that the sample might not be completely homogenous. If the two halves taken to separately measure potassium-40 and argon-40 contain different amounts of these isotopes, the date will be incorrect.

To get around these problems, an ingenious procedure was developed at the University of California at Berkeley in the 1960s: The sample is sent to a nuclear reactor, where it is bombarded with neutrons. The bombardment triggers many nuclear reactions in the sample, one of which turns a portion of its potassium into an isotope of argon, argon-39. Thus argon-39 becomes a proxy for the potassium content of the sample. Then, instead of measuring potassium-40 in one part of the sample and argon-40 in another, one need only measure the two argon isotopes in the neutron-irradiated sample—which can be done very precisely using a mass spectrometer. The potential uncertainties associated with weighing out the sample completely disappear because it is only necessary to measure the *ratio* of the two argon isotopes, which is independent of sample size.

This procedure, usually referred to as argon-argon dating, is now the preferred approach in potassium-argon age determinations because it leads to more accurate ages. However, because the nuclear reactors used for irradiating samples vary in their characteristics, each batch of samples must be accompanied by a standard rock sample with a well-known age (which is based on dates from several laboratories, using other techniques). Provided the results from unknown samples are calibrated to this standard, the dates measured by different research groups are strictly comparable, regardless of variables such as reactor conditions.

In the case of the glass spheres from the K-T boundary, the different research teams used slightly different values for the age of their standard when they did the age calculations. This resulted in small differences in the dates they reported. Furthermore, since the time of the original work in the 1990s, the age for that same standard has been updated, and a new value is now accepted internationally among geochronologists. If the original dates are recalculated using this new value, the results from different laboratories agree perfectly and give the now-agreed-on date of 65.5 million years.

Although the pathway to the correct age for the K-T boundary—including the use of standards—may seem a bit tortuous, it is important to understand because it illustrates the care necessary to get things right. When geochronologists today report the results of dating studies, they include every conceivable detail—the ages they adopted for various standards, the particulars of the equipment they used, the value they chose for the half-life in the calculations, and everything else that may be pertinent for understanding their results. That makes it possible to go back to the original data and recalculate the results if, for example, the age of the standard is updated, or a new and more accurate value for the half-life is determined.

The practice of using standards is a common one in many fields of science, not just geochronology. Although modern instruments are often very precise (in the sense that they give the same result when the same sample is analyzed multiple times in succession), different laboratories may obtain slightly different values for measurements of the same sample. Even within the same laboratory, results may vary slightly for analyses made months apart. There are many possible causes for such discrepancies (which are usually quite small). However, such problems can all be circumvented if the analyses are calibrated through frequent measurements of a standard. In some fields, standards are developed and certified in government laboratories, such as the National Institute of Standards and Technology in the United States. In others, standards may originate in a research laboratory as an in-house aid in analysis and eventually gain widespread recognition and use—as is the case for the standards used in potassium-argon dating. Whatever their origin, however, standards are invaluable for assessing whether results from different laboratories are reliable and comparable.

The K-T boundary is unique among the major time scale boundaries because it is marked worldwide by a recognizable layer of sediment that records a geologically instantaneous event. Even where this layer is not very obvious visually, it can be identified chemically by its characteristic iridium enrichment, which precisely identifies a sedimentary layer dating to 65.5 million years. This date establishes the exact timing of the major evolutionary changes that are evident from the very different fossil assemblages found on either side of the boundary.

Identifying and accurately dating places in the rock record where both major and minor biological changes took place is obviously important for understanding evolution. Paleontologists are often despondent about the fact that there is so much missing from the fossil record—it has been estimated that perhaps only 1 percent or less of all species that have ever lived have been found as fossils, and most never will be. This is partly because of the vagaries of preservation, but also—especially for the distant past—because previously existing sedimentary rocks have been destroyed by erosion or plate tectonics, or have been metamorphosed to the point where fossils are no longer recognizable. However, even from the tiny remaining fraction of fossilized species, it has been possible to construct an impressively detailed understanding of how evolution has operated, and what its time scale is.

The two tasks—paleontology on the one hand, and geochronology on the other—are often taken up by different groups of scientists, because most paleontologists are not themselves geochronologists, in spite of the crucial importance of time in their discipline. Sometimes researchers from the two disciplines collaborate on problems that seem especially interesting, or simply to work out a local geological problem. But, in general, the numerical and fossil-based aspects of the time scale have progressed in parallel. Today there is an international body, the International Commission on Stratigraphy, that has as a primary goal the integration and evaluation of such data. It seeks to unify the work of thousands of scientists across countries and continents, and periodically publishes a standard global geological time scale that includes details of the type localities for boundaries, and assessments of their most precise radiometric ages. In a very real sense, these scientists are carrying on the work of William Smith and the myriad geologists who followed him in working out a detailed picture of the history of life on Earth. The result—our current understanding of evolution and our planet's geological history—must truly rank among the most important accomplishments of modern science.

In the remainder of this chapter, I will try to summarize, necessarily briefly and somewhat selectively, some of the details of our current understanding of biological evolution, emphasizing the role that geochronology has played. That task begins with the question, When did life on Earth arise (and, of course, also, how did it appear, and what was it like)? There are not yet firm answers to these questions, but there are many clues. What are currently claimed to be the earliest signs of life—and this is still a controversial claim—are found in ancient rocks from western Greenland. The evidence is not in the form of recognizable fossils, but rather in the properties of small blebs of graphite—pure carbon—that occur in metamorphic rocks and are thought to have a biological origin. The Greenland rocks are so old that over their long history they have been buried, heated, folded, and recrystallized almost beyond recognition, so much so that it has been difficult to determine whether they were originally sedimentary or igneous rocks. But recent detailed investigations by a team from the University of California and the University of Colorado have demonstrated that the Greenland rocks were almost certainly precipitated chemically from an ancient ocean. That at least allows the possibility that the graphite is carbon from once-living organisms that lived in the sea.

Carbon is one of the most abundant elements in the universe and is essential for all living things. As explained earlier in this book, ordinary carbon is made up of two stable isotopes, carbon-12 and carbon-13; recall also that the numbers designating the isotopes are a measure of their mass, so carbon-12 is the "lighter" of the two isotopes. During biological processes such as photosynthesis, when living organisms take carbon from the environment to make the various components of their cells, they preferentially take up the lighter isotope because carbon-12 has slightly weaker chemical bonds than carbon-13, and it therefore reacts more readily. Thus biological carbon is always "light," enriched in carbon-12, an isotopic fingerprint that can be used to distinguish whether a particular sample is biological in origin. Even severe metamorphism that converts organic remains into graphite doesn't affect this signature. Just such a process apparently produced the Greenland graphite, because it carries the isotopic fingerprint of biological carbon: it is enriched in carbon-12.

The results for the Greenland samples are not in question: they indeed signify a biological origin. But some researchers have voiced doubts about whether this signature is original or was modified later (for example, by addition of organic carbon long after the rocks formed). There are passionate advocates on both sides of this controversy, but, as in all scientific debates, the matter will eventually be settled through continued research and collection of evidence. If for the moment we accept that the biological fingerprint is original, we can ask when the organisms that were eventually turned into graphite actually lived. In other words, what is the age of the original sedimentary rocks?

That question has now been answered definitively. The same team that confirmed the sedimentary origin of the graphite-containing rocks also identified bands of igneous rocks that cut across—and are therefore younger than—the sedimentary units. Zircon crystals extracted from these igneous rocks have been dated using the uranium-lead method, and they give an age of 3,825 million, or 3.825 billion, years. The graphite-bearing sediments must be older, although it is not possible to say exactly how much older.

If the Greenland graphite really did originate as biological carbon, life existed in the Earth's oceans before 3.8 billion years ago, only some

700 million years after the Earth formed. In some ways, this is not so surprising. Cesare Emiliani, an isotope geochemist who studied under Harold Urey during the heady days of discovery at the University of Chicago and who has written widely about the Earth, made a perceptive comment about early life on Earth. Some people claim that life is a miracle, he said, but what would really be a miracle would be the *absence* of life on Earth. All the necessary conditions are in place, and have been so from very early in our planet's history. We are the right distance from our principal source of energy, the sun, and the Earth hosts all the necessary chemical elements and compounds. Laboratory experiments (the first ones also done at the University of Chicago, by one of Harold Urey's students) have shown that amino acids, simple organic compounds but important ones for life, form readily under conditions thought to characterize the early Earth. Time was no problem—over millions or tens of millions of years, almost every conceivable chemical reaction that could have taken place among the existing compounds must have occurred. In this view, it is no surprise that the 3.8-billion-year-old Greenland rocks contain evidence for life. Some researchers have suggested that life may have arisen considerably earlier and possibly even more than once, only to be wiped out by the continual early bombardment of the Earth by objects many times the size of the asteroid that struck at the end of the Mesozoic era.

Unfortunately, carbon isotopes can't tell us anything about the nature of early life, only that it existed. However, it is presumed that the organic carbon—now graphite—in the Greenland rocks came from prokaryotes (from the Greek *pro*, "before," and *karyote*, "kernel" or "nut"); the prokaryotes are single-celled bacteria with no nucleus in their cells. The later eukaryote ("true kernel") cells contain nuclei and other internal features. Prokaryotes are the first living things to be seen as true fossils in sedimentary rocks, and in numerical terms they still dominate our world. In some ways, they are the Earth's most successful organisms. They fill every conceivable ecological niche in vast numbers; there are more bacteria in your stomach than there are human beings on the planet. Unlike many other living creatures, prokaryotes seem to be in no danger of extermination due to human activity.

The oldest features believed to be fossil prokaryote cells appear in 3.5-billion-year-old sedimentary rocks from Africa and Australia. They are tiny—they can only be studied with a microscope—and, as with the Greenland graphite, there is controversy about whether they are truly biological in origin. Geologist William Schopf of the University of California at Los Angeles, one of those who first described the putative fossils, has identified eleven different "species" on the basis of slight differences in their morphologies. Most have the appearance of a series of cells joined together in curving, filament-like strands. Some of the fossil objects look very much like present-day cyanobacteria (often referred to as blue-green algae), which produce oxygen through photosynthesis. Cyanobacteria are found in most bodies of water—they sometimes form "blooms" during warm summer months, making the water murky and leaving a layer of organic material on the surface. Taking his cue from the similarity between his fossil organisms and today's cyanobacteria, Schopf says that "pond scum" has been the dominant life form for four-fifths of the Earth's history.

It is difficult to prove beyond reasonable doubt that the microscopic features really are fossils of once-living organisms. Schopf himself set out a series of criteria that must be met, including some that are geological (for example, proof that the features are indigenous and at least as old as the rocks in which they are encased) and some that are biological. The latter are the most difficult to assess, because they rely primarily on the similarity of the microfossil shapes and sizes to well-documented fossil organisms or to present-day living organisms. Schopf is confident that his examples meet the criteria he has set, but some other researchers who have independently studied these microfeatures from the same deposit have challenged his interpretation. They reinterpreted both the geological setting of the site in western Australia where Schopf had collected his samples and the microfossils themselves, and proposed that the microfossils formed nonbiologically in ancient veins through which hot, metal-rich solutions flowed.

As this is written, the debate about the Australian microfossils has not been settled conclusively. But if Schopf is correct, prokaryotes were thriving 3.5 billion years ago, and maybe earlier. The age of his samples, 3,465 million years, has been determined by uranium-lead dating of zircons from igneous rock units that bracket the layers containing the disputed microfossils.

The difficulties inherent in detecting evidence for life in ancient rocks are enormous. You can gain some understanding of the problem by looking at the palm of your hand. There are thousands of bacteria there, but you can't see them because they are microscopic. Now imagine trying to find the fossilized remains of such organisms in heavily metamorphosed, 3.5- or 3.8-billion-year-old rocks. Most biologists and geologists are convinced, however, from various strands of evidence, that life arose "very early" and that single-celled bacteria were the very first organisms to populate our planet. Most also agree that there was probably life on Earth 3.8 billion years ago. But not all agree that the Greenland graphite or Schopf's microfossils are really evidence of that life. Their skepticism drives the search for new ways to tackle the question. One that has proved successful is the use of "chemical fossils" that—like the isotopes of the Greenland graphite—in some way reflect biological activity.

Geochemists refer to organic compounds that are specific to certain types of organisms as biomarkers. Recent studies of organic-matter-rich shale from Australia show that it contains biomarkers specific to cyanobacteria, and also biomarkers characteristic of the more complex eukaryotes. Eukaryotes, with nuclei and other internal structures in their cells, are slightly more "advanced" than the prokaryotes and are thought to have evolved from them (and therefore to have appeared on Earth later). So far, the oldest known rocks that contain unambiguous fossils of these eukaryotes are about 2 billion years old. However, the Australian shales containing eukaryote-specific biomarkers have been dated to 2.7 billion years, greatly expanding their known history. This discovery also implies that their ancestor prokaryotes, organisms like William Schopf's pond scum, existed much earlier than that.

The search for fossil evidence of very ancient bacteria recently took yet another and a quite unexpected turn when researchers discovered that even igneous rocks might harbor indications of life—something almost nobody anticipated. Sedimentary rocks had always been the focus; igneous rocks were supposed to be biologically barren. But bacteria are far from being gourmets—they have an amazing capability to "eat" almost anything. Some varieties will gobble up plastic, others oil spills. Bacteria have been touted as possible solutions to cleaning up toxic waste—throw in a handful of bacteria and let them eat and multiply. And some, it now appears, eat solid rock. They have been found deep underground on land and, of particular interest, colonizing the natural volcanic glass that is ubiquitous along the great undersea midocean ridges. The ridges are sites of frequent volcanic activity, and, when fresh lava flows out onto the sea floor, it instantly freezes on contact with cold seawater, forming natural glass. The bacteria actively dissolve the glass to obtain the chemical elements they use as a source of energy, and, in so doing, they leave behind tiny but deeply penetrating (in microscopic terms) tubules. These features have been found in literally hundreds of different samples from all of the world's ocean basins, in lava flows with ages ranging from essentially zero to 145 million years. If bacteria existed in the oceans billions of years ago, similar evidence of their presence should occur in ancient rocks, too.

Recently, a multidisciplinary group of researchers from Canada, Norway, the United States, and South Africa took up this challenge and began looking for the microscopic tubules in ancient rocks. By 2006, these scientists had found them in several samples from South Africa. Not only do the tubules resemble the modern varieties in shape and size, but they also exhibit other (chemical) evidence for biological activity. The South African rocks have well-preserved features that mark them as lava flows erupted onto the sea floor. Their age has been determined by two measurements, one a potassium-argon result for minerals of the lava itself, the other a uranium-lead date on zircons from an ash layer interbedded with the lava flows. Within the experimental uncertainties, the dates from these analyses are the same: 3.48 billion years.

Thus, even if the Schopf fossils are not really fossils at all, there are good reasons to believe that our world was full of bacteria by 3.5 billion years ago, and probably much earlier. But, for a very long time after that, there was little else—at least little else that left a fossil record. As we have already seen, there were eukaryotes by 2.7 billion years ago, and there are also fossils of multicellular plants in the form of simple algae. But there is no record of multicellular animals until approximately 600 million years ago (see figure 20), when the complex organisms of the Ediacaran period appeared.

At first these animals were rare and simple, but they quickly increased in both abundance and diversity. Ediacaran fossils appear not long after the end of a series of global ice ages that are referred to as "Snowball Earth" episodes because of their extent (glaciers covered the land even into the tropics) and severity. This may or may not be a coincidence; the rapid burst of evolution that characterizes the Ediacaran animals has been ascribed by some paleontologists to the worldwide return of warm, equable conditions after a prolonged cold interval. The last of the glacial sediments from the Snowball Earth period date to approximately 595 million years ago. Accurate dates for the Ediacaran organisms are sparse. However, in a few localities, there are volcanic ash layers within the sedimentary rocks hosting the fossils. The oldest reliable age, 565 million years, comes from uranium-lead dating of zircons from one such ash layer in eastern Newfoundland, Canada, where the assemblage of Ediacaran fossils depicts a flourishing and varied fauna. Thus, after a very long period—perhaps 2 billion years—of apparent stability and little change, in something less than 30 million years, a diverse and complex group of animals evolved. By the beginning of the Cambrian period, 542 million years ago, they were gone.

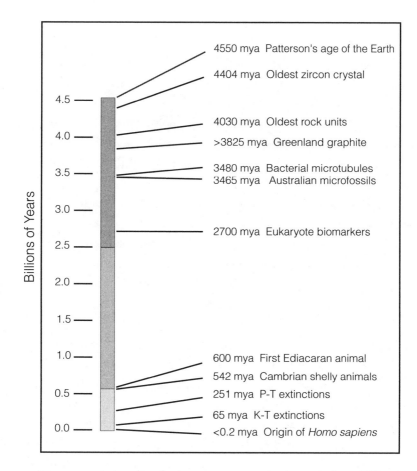

Figure 20. A time line for important events in the evolution of life on Earth. The scale on the left is in billions of years ago; event ages are given in millions of years ago (mya). Shading on the vertical bar shows four major divisions of geological time: *from top to bottom,* the Hadean (from the Earth's formation to 3.8 billion years ago); the Archean (from 3.8 to 2.5 billion years ago); the Proterozoic (between 2.5 billion years ago and the beginning of the Cambrian period 543 million years ago); and the Phanerozoic (from the beginning of the Cambrian to the present).

Some researchers think that the boundary between the Ediacaran and Cambrian periods marks another of the periodic mass extinction events that characterize the geological record, the result of some form of global environmental change. Others think the Ediacaran extinctions were caused by the new kids on the block, the animals of the "Cambrian explosion" (the descriptive term paleontologists use for the very rapid proliferation of body forms that occurred at this time). To the soft and unprotected Ediacaran animals, these new creatures with skeletons, shells, and chitinous armor must have been voracious predators. But that still does not answer the question, Why did the new animals appear in the first place?

When the geological time scale was first developed in the nineteenth century, the small, often indistinct traces of the Ediacaran fauna had not even been noticed. The shells and skeletons of Cambrian and later animals, however, left obvious, well-preserved fossils. To early geologists, it was as though the Cambrian animals had magically appeared, suddenly abundant in number and—a key observation—body form. Biologists classify animals on the basis of body form, and by partway through the Cambrian period all the varieties recognized in animals today were present in the fossils.

It has always been clear that the time scale for this diversification was short, but until recently nobody really knew how short. Once again, zircon crystals in volcanic ash layers came to the rescue. In an elegant study of Cambrian rocks from Siberia, published in *Science* in 1993, Sam Bowring and a group of Russian, MIT, and Harvard collaborators were able to show that it all happened in a geological blink of an eye.

Paleontologists have focused much attention on Cambrian fossils because of their evolutionary importance—they preserve a record of the diversification of animal life that led, eventually, to all the creatures we know today. Turnover was rapid, which has allowed researchers to partition the Cambrian period into a series of subdivisions based on the appearance and disappearance of various species, and this relative time scale makes it clear that the "explosion" in evolution took place partway through the period, not at the very beginning. The exact timing, however, was unknown until recently, because few sections of Cambrian sedimentary rocks were known to contain volcanic material suitable for dating. But the Siberian rocks—which are in an area that had been closed to foreign researchers during the Cold War because of a nearby Russian radar defense system—do. Still, the ash layers are few in number, and Bowring and his colleagues were able to find datable zircons in only a few horizons. The precise uranium-lead ages they measured, however, shrank the allowable duration of the Cambrian explosion to only 5 or 6 million years. For biologists and paleontologists, who have referred to the Cambrian explosion as the Big Bang of evolution because nothing like it is seen before or since in the geological record, this is remarkable. The pace of evolution was frenetic. By some criteria, there were more existing animal phyla (the basic subdivisions of biological organisms) in the Cambrian period than today. The speeded-up chronology revealed by the dating has prompted even more intensive study of the Cambrian explosion in an effort to understand its underlying causes. No definitive answer has yet emerged, but when it does, it is quite likely to be as startling as the fact of the explosion itself.

From the time of the Cambrian explosion until today, we have a more-or-less continuous fossil record of life on Earth, although it represents only a small fraction of the species that have existed and is biased toward those that lived in the oceans, because they are most readily preserved in sedimentary rocks. Within that fossil record, dating studies have shown (as we have already seen) that the duration of extinctions at both major boundaries—the K-T and P-T—was very short, measured in hundreds of thousands of years, not millions. In both cases, the bulk of the extinctions appears to have taken place entirely within the range of uncertainty that characterizes the date for the boundary itself—the timing can't be pinned down any more precisely than that. This is perhaps understandable if a major cause is impact, as seems to be true for the K-T boundary. But there is no definitive evidence for a similar event at the P-T boundary (although scientists have searched for clues assiduously). What, then, could have had such catastrophic global consequences that almost all life on Earth was wiped out?

One possibility has gained credibility in recent years on the basis of careful dating studies. Potassium-argon ages for one of the largest outpourings of volcanic rocks in the Earth's history, the Siberian flood basalts, completely overlap the precise dates for the P-T boundary at the type section in China. The so-called flood basalts are the product of anomalously large and often long-lived volcanic eruptions that "flood" the landscape with layer upon layer of lava. They occur in several areas of the world, notably western India, eastern South America, and, on a slightly smaller scale, in Oregon in the United States. The Siberian episode was so large that it would have injected huge amounts of volcanic gases (especially carbon dioxide and sulfur dioxide) into the atmosphere, with potential global consequences. Nevertheless, large volcanic events, however rapid, cannot match the instantaneous effects of an impact, and whether—or how—the Siberian flood basalts link to the P-T extinctions remains unclear. But the exact coincidence of ages is compelling. And the finding that both the K-T and P-T mass extinctions happened very quickly has strongly influenced ideas about how evolution proceeds.

Both these major mass extinctions were followed by rapid "radiation," a proliferation of new species from the small base of the few that made it through the crises. In such circumstances, "natural selection" would be expected to act in unpredictable ways, and random branching of evolutionary pathways would be the rule rather than the exception. Organisms would not be successful through advantageous adaptation to a stable or slowly changing environment; rather they would be likely to survive the abrupt crisis because of some chance trait, while "fitter" creatures perished. "Survival of the fittest" would take on an entirely different meaning. Some researchers have suggested that this is exactly how the rise of the mammals—leading eventually to you and me—occurred. For nearly 100 million years, mammals coexisted with the dominant dinosaurs, but they remained small in size and were of secondary importance among the vertebrates. Had an asteroid not struck 65.5 million years ago, that might still be the case. However, although small, the mammals were relatively abundant, inhabited many different habitat types, and had varied diets. These traits allowed at least some of them to survive the crisis while the much larger and more specialized dinosaurs perished. In a very short time after the extinctions, diverse mammal groups colonized both land and sea. Similar examples undoubtedly exist for the earlier P-T extinction and subsequent radiation, although because of their antiquity they are not so easy to trace.

Ultimately, evolution is a molecular affair, the result of mutations and other changes in genes (primarily in DNA and RNA) that get passed on first when cells divide, and then from generation to generation. Genetic information determines every aspect of living organisms. Some mutations make no difference whatsoever to succeeding generations, but others—especially in the face of external pressure—lead to varying degrees of success or failure among descendants. Genetics determined which organisms made it through the P-T and K-T extinctions, and also guided the biological radiation that followed.

It is natural that we humans are fascinated by the genetic pathways of our own past. The revolution in biology brought about by molecular genetics has shown beyond any doubt that every single living thing, from a monarch butterfly to a grizzly bear or a United States senator, shares a common ancestry. But it is the more recent connections to other species that have captured most attention. From the time Darwin concluded that we are closely related to African apes, researchers have attempted to work out the details of this lineage. The fossil record is sparse because many of the apes lived in tropical, wooded environments where fossils are rarely preserved. However, molecular genetics has confirmed Darwin's conclusion; we are indeed part of the biological family known as the "great apes." Our closest living relatives within that family are the chimpanzees, and the genetic studies show that first the gorillas, then the chimpanzees, branched off from our line of descent. (Of course, this is an anthropocentric view of the process; the chimps would contend that we split off from their lineage.) The exact timing is difficult to determine, but by making some assumptions about the rate at which genetic change occurs, researchers have estimated that the gorillas branched off around 6 to 8 million years ago, and the chimpanzees at 4 to 6 million years.

Tracing evolutionary changes gets easier after the divergence of the chimpanzees because more abundant and more reliable radiometric dates are available. For this we can thank our distant ancestors for living in eastern Africa—if they had lived in, say, Kansas, we would have much less information about the time scale of their evolution. The Rift Valley of eastern Africa is a place of active volcanism because, as the name implies, it is a place where the continent is rifting apart, causing hot magma to well up from the interior. As a result, our predecessors were well acquainted with volcanic eruptions, which frequently spread volcanic ash throughout their habitat. The ash is interlayered with fossils, and, in at least one place—Laetoli, in northern Tanzania— it has even preserved the footprints of our ancient ancestors, pressed into the soft, damp ash, a record of their passing as they went about their daily lives. But the most important aspect of the ash layers is that they provide datable material, and recognizable time horizons spread over large areas. The footprint-bearing ash, for example, has been dated at 3.6 million years.

The method of choice for dating the East African ash layers is the potassium-argon technique. Some of the layers contain zircon crystals, but uranium-lead dating is simply not feasible for grains that are only a few million years old. The potassium-rich minerals in the East African volcanic deposits, however, have accumulated enough argon-40 for straightforward potassium-argon analysis.

The fossil record of hominids is fragmentary, and the recovery of even moderately complete specimens is quite rare (hominids are members of the biological family of "great apes," which includes *Homo sapiens* and various now-extinct precursors). In spite of that, through collaboration between paleontologists who focus on anatomical details and geochronologists who continually work to improve their dating methods, a fairly detailed time line for human evolution has emerged. Debate continues about the exact pathways, but all researchers recognize the importance of three different genera, each with a number of species. The oldest and most primitive is *Ardipithecus,* which appears in the fossil record almost 6 million years ago (see figure 21), very close to the time when molecular genetics data suggest that hominids and chimpanzees split off from their common ancestor. *Ardipithecus* was a small, chimpanzee-like animal that may have walked upright, and that had sharply pointed canine teeth, in contrast to those of more modern hominids. Next along the lineage came the larger *Australopithecus,* generally considered to be the immediate predecessor of humans. Finally, about 2.5 million years ago, *Homo,* our own genus, appears. Fossils of various species from all three of these genera occur in eastern Africa, and throughout the entire time span of their existence there are volcanic ash layers that provide a framework of reliable ages.

The first inkling that our hominid ancestors had roamed the East African countryside millions rather than hundreds of thousands of years ago came in 1959, when British archaeologist Mary Leakey, working together with her husband, Louis, at the Olduvai Gorge of northern Tanzania, discovered a hominid skull. By coincidence, Jack Evernden, a geophysicist from the University of California at Berkeley, who was one of the early developers of potassium-argon dating, had visited the site two years earlier and collected samples of volcanic ash for dating. At a time when other laboratories were focusing on applying the then-new dating method to very ancient samples, Evernden and his colleagues wanted to tackle the young end of the time scale. The East African ash layers seemed to be a good target, especially because they contained minerals that were very rich in potassium. Evernden didn't have any great expectations for the Olduvai samples; at the time, no hominid remains had been found there. On the basis of fossils in adjacent sediments, the Leakeys thought the ash layers might be as much as a few hundred thousand years old, but the estimate was very rough.

The 1959 discovery made Olduvai Gorge and the Leakeys famous.

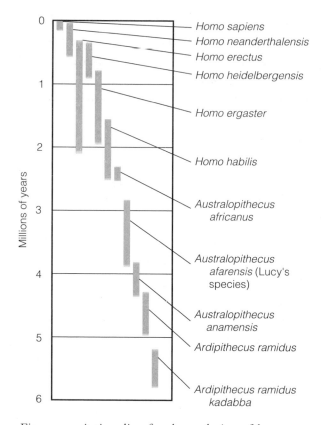

Figure 21. A time line for the evolution of humans. Each vertical bar shows the approximate lifetime for a species; origination and extinction times are not known precisely because they depend on the availability of datable fossils.

Louis Leakey, who believed that the skull was from a direct ancestor to *Homo sapiens,* proposed a new genus for the fossil, *Zinjanthropus,* after the ancient word for the east of Africa, Zinj. The fossil quickly became known familiarly as Zinj. But what really hit the headlines was the potassium-argon date. The volcanic samples collected by Evernden had already been dated in Berkeley, but it wasn't immediately known where they came from relative to the level where Zinj had been found. After some hurried consultations, that problem was worked out. The potassium-argon results indicated that Zinj was 1.75 million years old. This was far older than anyone, including the archaeologists working on hominid fossils, had ever suspected, and it changed their thinking about hominid evolution. Suddenly, anthropologists, paleontologists, and archaeologists wanted to learn more about the new dating method geologists were developing.

The Olduvai Gorge fossil has now been reclassified as belonging to the genus *Australopithecus,* and it is no longer considered to be a direct ancestor of modern humans. But in 1974, fifteen years after Zinj was discovered, another *Australopithecus* fossil was found that may be. This one, named Lucy, also made international headlines. Lucy's fame came because of the unusual completeness of her skeleton and her great age, but it didn't hurt that her name was instantly recognizable from the Beatles' famous song "Lucy in the Sky with Diamonds."

Lucy's discovery was serendipitous. Donald Johanson, now director of the Institute of Human Origins at Arizona State University, was heading back to his field camp in Ethiopia and happened to see a bone sticking out of a layer of sediments. Like any respectable anthropologist, he stopped to take a closer look—and immediately recognized it as a hominid arm bone. That was a bit like finding a gold nugget lying on the beach, and on close examination it got even better. There were a lot of other bones around, including part of a jaw.

Johanson was spending his second field season in the Afar region of Ethiopia. The now-legendary Maurice Taieb, a Tunisian-born geologist who had earned his doctorate at the University of Paris, had organized both expeditions. His PhD thesis was based on field studies in the Afar, and, with many years of mapping and detailed geological investigations under his belt, Taieb believed the region had great potential for study-ing human ancestry. He was right. The Afar area has turned out to be perhaps the most prolific place in the world for hominid fossils. Even so, they are not exactly lying around everywhere. But during the rains, when water courses through the gullies and washes, the friable sedimentary rocks erode easily. Each year, new material is uncovered; occasionally a hominid fossil or two is exposed.

Lucy was discovered near the northern end of the East African Rift Valley, close to the Red Sea. When she was alive, the regional climate was much wetter, and the area hosted large lakes. Most of the sedimentary rocks that contain hominid fossils were deposited within or along the margins of those lakes, or along the streams that fed them. It was an ideal environment for preservation of remains, and in addition to the hominids there are fossils of a multitude of other animals that lived side by side with these ancestral humans.

It took several weeks of careful work to remove all of Lucy's bones from the sediment in which she was entombed. Although the fossil was called Lucy from the very beginning, it was not even clear initially that she was female. However, the consensus now is that she was indeed a woman. All other known fossils of Lucy's species from the Afar region fall into two distinct size groups—larger males and smaller females. Lucy fits neatly into the smaller group, standing about three feet eight inches high and weighing only sixty to sixty-five pounds. Her skeleton, estimated to be about 40 percent complete (an impressive figure if you consider that most hominid finds comprise only a few teeth, or one or two bones), is now housed in the National Museum of Ethiopia in Addis Ababa.

Although word of Lucy's discovery traveled quickly through the scientific grapevine, the first formal description came in a paper in *Nature* in March 1976. It listed all finds from the 1973 and 1974 expeditions to Ethiopia, most with terse descriptions such as "left lower molar," or "right proximal tibia," or "right distal femur." Lucy was described as a "partial skeleton." But her popular name didn't appear in the *Nature* paper; she was identified only as specimen AL 288–1.

Her age didn't appear either, at least not her precise age. She was "estimated" to be about 3 million years old, based on potassium-argon dates for basalt flows and volcanic glass shards from ash layers near where she had been found. The ages were imprecise, however, and were not consistent with one another. For a while, Johanson was fond of saying at meetings that he "was still dating Lucy." Clearly, more detailed age determinations were needed to put Lucy and other hominid remains from the site in their proper temporal framework.

But political events intervened. It became very difficult for foreign scientists to get permission to work in Ethiopia, and it was also increasingly dangerous to travel in the field. Warring tribesmen, it seemed, were everywhere. It was not a place for strangers; local interpreters and guards were just as likely to flee when confronted by rival tribesmen as to try to help out the archaeologists and paleontologists. However, the long gap had its benefits. By the 1990s, when it became somewhat easier, and—marginally—safer to work in the Afar region, great strides had been made in potassium-argon dating. Argon-argon dating, the improved variant of potassium-argon dating described earlier in this chapter, had become commonplace. And the mass spectrometers used to measure argon isotopes had become much more efficient and capable of measuring very small quantities of the gas. Furthermore, a method had been developed that allowed scientists to extract argon from a single crystal the size of a grain of sand, using a laser.

One of the reasons dates for volcanic ash layers had sometimes been inconsistent, with different samples giving slightly different ages, is obvious enough if you think about it. A volcanic eruption is a messy thing. Fresh lava gets sprayed up into the atmosphere and falls to form an ash layer, but so, too, do pieces of the volcano itself, entrained in the erupting material. Some of those pieces might be from much earlier eruptions. A typical sample for conventional potassium-argon dating may contain hundreds, or even thousands of individual grains, and, even if only a few of these are from previous eruptions, the "age" of the overall sample would be incorrect, skewed by the older grains. This became quite clear when the laser technique was put into practice and multiple crystals from the same ash layer were analyzed, one by one. Most grains would give the same result, but invariably there would be a few outliers, nearly always considerably older than the rest. By discarding the older dates, the correct age could be calculated with a high degree of confidence. Furthermore, in an environment like that of the East African Rift Valley, fossils are not always found in their original positions. Bones and even more-or-less complete skeletons can be washed away by flowing water and deposited in a low-lying depression. So, too, can the loose components of an ash layer. In the process, "foreign" grains can be mixed into the layer containing the fossil. Usually, grain-by-grain dating can distinguish these interlopers.

The laser method provided the first accurate age for Lucy, which was published in the journal *Geology* in 1994. A prominent volcanic ash layer lies just below the level at which Lucy's skeleton was found, and careful analysis of multiple grains date it to 3.18 million years. Lucy herself probably did not witness the eruption in which those grains were formed, but undoubtedly some of her not-too-distant ancestors did.

When Lucy was found in the 1970s, some thought that she might be a "missing link" between more apelike creatures and ourselves. After all, even without precise dating, it was clear that she was very old, and her skeleton showed that she had an upright posture. But now, with many more hominid fossils discovered, and with most of them placed in an accurate time framework, the picture is clearer—and at the same time, more complicated. Lucy is actually one of the more "advanced" of the Australopithecines (see figure 21). The first known *Australopithecus* fossils date from 4.2 million years, a million years before Lucy appeared on the scene. And the last of Lucy's species died out, as far as we know from the fossil record, several hundred thousand years after Lucy herself, at around 2.8 million years ago.

Although she is no longer considered to be a missing link, Lucy may indeed be part of the direct lineage that leads to humans. Anthropologists still debate that possibility, sometimes heatedly, because, many say, the fossil record is so poor in hominid remains that we simply can't tell. But it is true that the first known examples of our genus, *Homo,* date to 2.5 million years, only shortly (in geological terms) after the last of the Australopithecines disappeared. *Homo habilis,* as that early species is

known, had a bigger brain than Lucy and, for the first time in the hominid fossil record, used simple stone tools. *Homo habilis* also occurs in
eastern Africa. As there appears to be no overlap between Lucy's species
and *Homo habilis,* it is quite possible that one is the direct descendant of
the other.

Within our own genus, *Homo,* there are many recognized species, but
only one living example, *Homo sapiens* (which is probably just as well,
given the difficulties that arise even among groups within a single
species). As has been done for the other hominid genera, a temporal
framework has been built up for *Homo* on the basis of careful radiometric dating, primarily using the potassium-argon method, coupled
with anatomical studies (see figure 21). One of *Homo*'s prominent features is long periods of overlap between some species. This suggests
there was considerable divergence among the different species of *Homo,*
and probably also significant competition. It also makes it difficult, with
present information, to discern our own lineage exactly. But the sediments of eastern Africa may hold still more clues to that puzzle. If they
do, you can be sure that precise radiometric dating of the interlayered
volcanic ash horizons will play a vital part in deciphering them.

CHAPTER EIGHT

Ghostly Forests and Mediterranean Volcanoes

Most geologists are historians—not historians as we usually think of them, but historians of the Earth. They talk about the Cretaceous or the Precambrian as easily as conventional historians discuss the sixteenth century. And, like conventional historians, geologists tend to have specialties. Some are experts in what was going on during the Jurassic period; others spend their careers examining trilobites from the Cambrian or metamorphic rocks from the Archean. But all historians, regardless of what slice of time they study, need a time line to order events and fit them into the wider historical picture. For geologists, the particular slice of time they work in often determines which of the radiometric dating methods will be most useful for their research. For those interested in the very recent past, the choice is severely restricted. Uranium-lead dating simply doesn't work because the uranium half-lives are too great. Potassium-argon dating, in favorable cases, can be pushed into the range of 10,000 to a few tens of thousands of years, but the uncertainties increase significantly at these young ages. For many problems in the zero to 50,000-year range, radiocarbon dating is the best choice, and sometimes the only viable option.

For Brian Atwater of the United States Geological Survey, radiocarbon dating does just nicely. Atwater was chosen as one of *Time* magazine's top one hundred influential people for 2005, in the Scientist and Thinkers category, and in a moment we will see why. Whether he enjoyed sharing the honor with Karl Rove only he can say. Atwater is definitely a long-term thinker.

What brought Atwater national and international attention was his discovery that there have been a series of huge earthquakes and tsunamis along the Pacific Northwest coast of North America in the not-too-distant past. At first, though, Atwater wasn't sure exactly when in the past they had occurred. He was faced with the question, How do you date an earthquake? Ever ingenious, Atwater—and others since—found ways to do this with radiocarbon, by dating once-living things, mainly plants, that had been affected by the earthquakes. Their research has shown that "great" earthquakes (those larger than about 8.0 on the Richter intensity scale) have shaken the region, on average, every 400 to 500 years. The most recent such event was in 1700. That is long enough ago that there were no European settlers on the West Coast to experience it, and there are no written records. Finding out exactly when it happened, even with the help of radiocarbon dating, required some coordinated detective work.

Most earthquakes occur along the boundaries between the tectonic plates that make up the Earth's surface. These plates, which can be many tens to more than a hundred miles thick, move around relative to one another in slow and sometimes deadly motion, and in places—the so-called subduction zones—one tectonic plate slides under another and down into the Earth's interior. Most of the great earthquakes occur along subduction zones. Because they are usually located near an ocean-continent border, many subduction-zone earthquakes occur underwater and generate large tsunamis, as happened in the great Indonesian earthquake of December 26, 2004. The energy released in such events is massive. The Indonesian earthquake caused the whole planet to shake, and although you couldn't actually feel the motion if you were thousands of miles away, the Earth's crust moved up and down by at least a fraction of an inch everywhere in the world. The United States Geological Survey calculated its magnitude to be 9.1 on the Richter scale. Translated, that means that in the space of just a few minutes, from a small geographical region, came a burst of energy roughly equivalent to the total U.S. energy use for an entire week.

Brian Atwater lives in Seattle, and he has a subduction zone almost on his doorstep. It is known as the Cascadia subduction zone, and it lies just off the Pacific Northwest coast, stretching from northern California to Vancouver Island, off the coast of British Columbia. By global standards, it is fairly short—much shorter than the similar features that extend all along the west coast of South America, or that curve around the south coast of Alaska and the Aleutian Islands. At the Cascadia subduction zone, one tectonic plate carrying part of the Pacific Ocean floor plunges under the plate that carries the whole of North America. The convergence goes on at the stately speed of about an inch and a half a year, which doesn't sound like much, but try multiplying by a few centuries of motion and it suddenly becomes quite significant.

Plates at subduction zones don't simply slide by one another continuously; they tend to lock up, stick for a while, and then—when enough stress has built up—they slip. The stress of years or centuries is released in an instant, and anyone or anything nearby gets thoroughly shaken by the ensuing earthquake. Most subduction zones experience frequent small earthquakes and periodic large ones. The Cascadia subduction zone, however, is an anomaly—it is, in terms of earthquakes, the quietest in the world. Some earthquakes do occur, but they are nearly all so small that they are never felt by the local populace. We know about them only because they are detected by sensitive seismometers. American historical records don't document any really large earthquakes along this zone either, which has led some scientists to suggest that a peculiarity of its behavior must prevent them from occurring.

However, Atwater knew that every other subduction zone has experienced great earthquakes. Two of the largest ever recorded had happened in his lifetime—in Chile in 1960 (9.5 on the Richter scale) and in Alaska in 1964 (9.2 on the Richter scale). Could it be that the apparent

lack of great earthquakes in the Pacific Northwest was simply an artifact, a consequence of the short span of written records, which went
back only a few hundred years? There were a few tantalizing clues in
Native American oral traditions, suggesting that large earthquakes had
struck before the Europeans arrived—tales of shaking ground, or of
tribes having to move because their land was abruptly flooded (possibly
because of coastal submergence during an earthquake, some geologists
thought)—but it was all very vague.

Atwater decided it would be useful to look at the geological evidence.
In both the Chilean and Alaskan earthquakes of the 1960s, low-lying
land along the coast had suddenly dropped, and the incursion of salt
water had killed off terrestrial vegetation and covered it with ocean
mud. Previously dry land was instantly turned into a tidal marsh. In
some places, there was evidence that this had happened repeatedly—
there would be a layer of soil with the remains of land plants, then a layer
of ocean mud, then another layer of soil, and so on. Apparently, after
each earthquake submerged the coastal land, mud and silt gradually
accumulated until it built back up to sea level. When the next great
earthquake struck, the land was once more submerged, starting the
cycle all over again.

Atwater began to investigate the bays and estuaries along the coast of
Washington State by canoe, in search of similar features. He soon found
what he was looking for. Just as in Alaska and Chile, there was evidence
for sudden submergence of coastal lowlands. In places, whole forests had
been drowned. In these "ghost forests," says Atwater, the trees "scream
at you." They are calling out for interpretation, he said: "How did I die?"
And the weathered, ghostly trees—now mostly just straight trunks with
few, if any, surviving branches (see figure 22)—are victims of only the
most recent submergence of the land. As he examined the muddy banks,
Atwater found a whole sequence of drowned horizons, suggesting that
the land had repeatedly and suddenly dropped relative to sea level. The
most logical explanation was that he was seeing the aftermath of numerous great earthquakes in the past.

Figure 22. A ghost forest at the mouth of the Copalis River, Washington State. The still-standing trunks are the remains of western red cedar trees killed by submergence accompanying the great earthquake of 1700. Spruce saplings can be seen growing near the tops of some of the dead trees. This photograph was taken by Brian Atwater in December 1997; since then, some of these trees have fallen over. Photo courtesy of Brian Atwater (this image appears in the book *The Orphan Tsunami of 1700*, published by the U.S. Geological Survey and the University of Washington Press in 2005).

In a few cases, the soil and drowned vegetation that Atwater found were covered with a layer of sand rather than fine mud, sometimes traceable over large distances. The sand layers always became thinner away from the water's edge. The coincidence of sudden submergence and deposition of a sand layer indicated to Atwater that he was seeing the combined effects of an earthquake and an associated tsunami—along these muddy shorelines, the only conceivable source of sand was offshore, and it could only be carried landward by the giant waves of a tsunami. Storm surges, even very large ones, would not be sufficient, and anyway it would be an unlikely coincidence for a great earthquake and a great storm to occur together—especially on several different occasions separated by hundreds of years.

The first results of Atwater's research were published in *Science* in 1987. He reported evidence for six abrupt submergence events along the coast of Washington State, each one probably caused by a great earthquake. However, he didn't have any information about their timing. The best he could do was to make an estimate—an ingenious one, but still an estimate. He used the fact that the repeated submergence and reemergence of the coastal land, involving elevation changes of just a few feet, could happen only if the average relative levels of land and sea in the area had been approximately constant. From other work, it was well known that this had been the case for about 7,000 years; before then, sea level had been lower, but rising rapidly due to the melting glaciers of the ice age. Atwater concluded that the six great earthquakes had occurred over the past 7,000 years—about one every millennium.

Suddenly, residents of cities like Vancouver, British Columbia; Seattle, Washington; and Portland, Oregon, began to worry. They didn't have a San Andreas fault in their backyard, and they were not used to being wakened by small earthquakes, as many Californians are. But they did have a subduction zone off the coast. If there had been at least six great earthquakes over the past 7,000 years, what were the chances of another one happening soon? That was an important question. To answer it required precise dating of the rapid subsidence events; if they occurred regularly, it might be possible to predict when the next one would strike. The presence of abundant organic material in the submerged horizons made radiocarbon dating an obvious choice for this work.

The ghost forests were tackled first. They were the most visible reminders of a past natural disaster, and they were also the most recent. In places, the drowned trees still stood tall, reaching thirty feet or more into the air (see figure 22). Counting tree rings might seem to be an obvious way to date these forests, but the tree trunks were heavily weathered, with most of the outer portions rotted away. This meant there were no samples available for accurate radiocarbon dating of wood from near the end of the trees' lives. However, by matching ring-width patterns from surviving portions of the trees with the known regional patterns, it was possible to estimate that the ghost forests had probably died sometime after about 1680. If a great earthquake was the cause, it had happened after that date.

Crucially, Atwater and his colleagues also found buried spruce stumps in the drowned forests. These had escaped serious degradation; their roots still had intact bark, and growth rings could be counted right up to the very last season of the trees' lives. The outer rings proved to be entirely normal in width, corroborating the conclusion that the trees had died suddenly. Although this didn't prove that the shoreline had been plunged below sea level during an earthquake, it was consistent with that scenario. The roots didn't have long sequences of rings that could be matched to regional patterns, but, by radiocarbon dating a sequence of rings and knowing that the outer ring marked the year of the earthquake, it should be possible to determine an exact age.

Atwater and his colleagues made radiocarbon age measurements on wood from nine different spruce stumps from two localities about thirty-five miles apart. They published their data in *Nature* in 1991, concluding that the drowning of the ghost forest had happened between 1695 and 1710. Even by the standards of the best radiocarbon dating studies, this was an amazingly precise result. If Atwater's group was right about the cause of submergence, the Pacific Northwest had been hit by a very large earthquake sometime during that fifteen-year interval.

How was it possible for Atwater and his colleagues to be so precise about the date? First, they had paid careful attention to all parts of the analytical procedure to minimize uncertainties. They had also analyzed nine different samples, and, by pooling the data, they were able to reduce the overall uncertainty below that of a single analysis. Lastly, the precision of their dates was due partly to the nature of the radiocarbon calibration curve in the age range they were dealing with.

Perhaps here it is worth flipping ahead to figure 25 on page 214 to remind yourself about this curve, discussed earlier in chapter 4. There I mentioned that radiocarbon researchers use the measured carbon-14 content of a sample to calculate its "radiocarbon age," which is offset from the true calendar year age. (The offset arises from the fact that, for consistency, the calculation is made assuming a constant value for the atmosphere's radiocarbon concentration, which is not really the case.) But the true age can be read from the appropriate portion of a calibration curve, such as the one shown in figure 25. You can see from this figure that, for steeply dipping sections of the curve, a fixed span of "radiocarbon years" will be read off as fewer calendar years, and that, for flatter portions of the curve, the radiocarbon years will correspond to more calendar years. Figure 25 doesn't extend up to the very recent past, but, if it did, it would show a very steep dip in the calibration curve between about A.D. 1600 and 1700. Knowing from the drowned forest tree rings that the submergence most likely happened after about 1680, Atwater and his colleagues typically counted back thirty to forty years from the outer ring before cutting out a wood sample for analysis. This, they thought, would probably put them in the steep part of the curve—and they were right. A twenty-year uncertainty in the "radiocarbon age" in this time interval translates to only ten or fewer calendar years. This further reduced the uncertainty of the age measurement.

The results of Atwater's work were widely disseminated, and, as usual, there were some skeptics—not about the date itself, which was generally agreed to be very sound, but about its implications. Perhaps there had been a series of earthquakes over a period of years; even the very precise radiocarbon age couldn't resolve this possibility. And, although it seemed clear that the event was "big," there was no way to estimate its magnitude accurately. Was it really a great earthquake? But then Kenji Satake, a seismologist with the Geological Survey of Japan, found the answer in an unlikely place: Japanese historical records.

Satake knew about the work in the Pacific Northwest, and he also discovered that there were historical reports in Japan of a large tsunami that occurred in January 1700. He wondered if there was a connection with Atwater's great earthquake. The timing, at least, was right.

Japan is no stranger to tsunamis, most of them generated by earthquakes that occur along its own offshore subduction zone. But events on the other side of the Pacific can also send giant waves crashing into Japan—the 1960 earthquake in Chile, for example, did just that, resulting in extensive damage and killing 140 people. There was no question that a great earthquake on the Cascadia subduction zone could cause a tsunami in Japan.

What caught Satake's attention in records of the 1700 tsunami was the absence of any mention of local ground shaking. This suggested that the source of the waves was distant. Satake could find no evidence in either historical or scientific writings for a large earthquake *anywhere* that was capable of generating a tsunami in Japan in 1700; Atwater's coastal submergence dated to 1695–1710 seemed to be the only match. Satake used the Japanese records to calculate just how big a Pacific Northwest earthquake would have to be to explain observed wave heights (which had been carefully recorded in the Japanese manuscripts he examined). He concluded that it must have had a magnitude of about 9 on the Richter scale—clearly a great earthquake. From the records of the waves' arrival in Japan, he was even able to pin down its timing: it had occurred at approximately 9 P.M. Pacific Standard Time on January 26, 1700.

There is something very satisfying about the combination of scientific results and historical sleuthing that made it possible to work out, to the hour, the time of an earthquake that occurred more than three hundred years ago and that had consequences on both sides of the Pacific Ocean. The story struck a chord with the public, too; when Satake's work was published in 1996, science writers from around the world picked it up, and their stories appeared in numerous newspapers and magazines. Interest in Cascadia earthquakes also spurred renewed investigations of references to earthquakes and tsunamis in the myths and oral traditions of native peoples from the Pacific Northwest. Roy Hyndman, a geophysicist at Canada's Pacific Geoscience Centre in Sidney, British Columbia, wrote a 1995 *Scientific American* article about Cascadia zone earthquakes and mentioned just such a story. The provincial archives in his hometown of Victoria, B.C., he said, hold oral history records from the original inhabitants of Vancouver Island telling of a large earthquake that struck the west coast of the island one winter's night. By the next morning, a native village at the head of a large bay had disappeared. It is just possible that this story—set in winter, and presumably referring to relatively recent history—documents the 1700 earthquake and tsunami. The Yurok people of coastal northern California similarly have myths that speak of shaking ground followed by flooding of the land. Such references are much vaguer than the written Japanese records, but they do show that native peoples of the Pacific coast experienced large earthquakes, subsidence of the shoreline, and, possibly, tsunamis.

When Atwater learned of Satake's work, he became so excited that he began to learn Japanese and arranged for a year's visit to Japan so he could examine the historical records himself. With the help of Satake and several other Japanese collaborators, he carried out a more detailed investigation of the archival material than had been done earlier, and, in so doing, considerably strengthened Satake's conclusion that a great Pacific Northwest earthquake was responsible for the January 1700 Japanese tsunami. The story of the detective work necessary to find and translate old documents is told in a book published by Atwater and his five collaborators in 2005, titled *The Orphan Tsunami of 1700*. In addition to helpful modern graphics, the book is beautifully illustrated with maps, pictures, and writings from shogun Japan.

Important as the work on the 1700 earthquake was, it documented only the most recent event. To get accurate information about the frequency of the great earthquakes, it would be necessary to date the sequence of older layers that Atwater believed also recorded submergence episodes. This turned out to be not quite as easy. If there had been ghost forests associated with these layers at one time, they had long since rotted away, removing the possibility of examining tree-ring patterns. Radiocarbon dates for the older layers had to be measured on small fragments of fossil plant material, such as twigs and leaves. Even with great care in sampling, it was always possible that "foreign" fragments that significantly predated the submergence could sneak into the samples. In addition, the many wiggles in the radiocarbon curve over the past several thousand years mean that there are quite a few intervals where even the most precise radiocarbon measurement translates only into a fairly imprecise calendar year age. Nevertheless, Atwater and his colleagues have now identified and accurately dated seven incidents of abrupt subsidence, beginning about 3,400 years ago and continuing up to the 1700 earthquake. This suggests, on average, a recurrence interval near 500 years. But the pattern is irregular—there is a gap of almost 1,000 years between 1,500 and 2,500 years ago, for example, and a cluster of three earthquakes between 1,000 and 1,600 years ago. That makes it difficult to predict exactly when the next one will happen. There is no doubt, however, that there will be a next one.

Atwater's pioneering work stimulated many others to search for new ways to shed light on Cascadia subduction zone earthquakes and tsunamis. One of the most interesting approaches was taken by Harvey Kelsey, of California's Humboldt State University, and his colleagues. These researchers found a small lake in southern Oregon that lies just over a quarter of a mile from the coast and that has been in existence for about 7,000 years. Bradley Lake, as it is called, first formed when shifting coastal sand dunes partly blocked the exit of a small stream to the sea, flooding the depression behind it. For most of its life, the lake has been just high enough above the high-tide level to be protected from storm surges—but not high enough to prevent large tsunami waves from rushing up the stream and dumping sand and salt water into it.

Over its lifetime, almost twenty feet of sediments have accumulated on the floor of Bradley Lake. Kelsey and his colleagues sampled these sediments by taking twenty-seven cores, carefully spaced out to cover most of the lake's area. When they began to examine them in the laboratory, they found that much of the sediment was, as expected, the product of the slow, monotonous, day-to-day and year-to-year rain of particles that characterizes all lakes. In places, they could even distinguish annual layers, with characteristically differently colored summer and winter sediments. But, periodically through the cores, they found evidence for catastrophic disturbances of this normal pattern.

In total, the researchers found signs of seventeen large disturbances, typically characterized by evidence for erosion—sometimes severe erosion—of the underlying sediments. The eroded horizons were usually covered by layers of sand and chaotic mixtures of lake-bottom mud, and this sequence of features could be correlated from core to core throughout the lake. Clearly the disturbances recorded major events. Fossils from below and above many of these disturbances showed the lake water had changed from fresh to brackish. This was sure evidence for the influx of seawater, and Kelsey and his colleagues concluded that the combination of sand layers and salt water must record large tsunamis. No other phenomenon could carry such a large amount of sand and seawater into the lake. In a few cases, the researchers were even able to distinguish successive waves from a single tsunami. The clue was disturbance intervals in which the sand comprised several individual layers, each with coarse sand at the bottom and finer sand toward the top. That's exactly what you would expect if successive slugs of sand were carried into the lake by successive tsunami waves—in each case, the coarser grains would settle to the bottom first, followed by the finer ones.

Kelsey and his colleagues calculated that any tsunami powerful enough to deliver sand and seawater to the lake had to originate from a Pacific Northwest earthquake—one in Alaska or Japan, even a very powerful one, simply could not produce such large tsunami waves. And the small number (four) of disturbances that didn't show evidence for saltwater incursion, they concluded, must be due to earthquake shaking that simply dislodged material from the lake margins and dumped it onto the lake floor but which did not generate a large tsunami.

Dating these events was one of the most important parts of the research, but also one of the trickiest. To begin with, the sand layers didn't contain any material suitable for radiocarbon dating. But there was a great deal of organic debris in the chaotic layers of muddy sediments that marked each disturbance. The sloshing around of lake water, whether from an earthquake, a tsunami wave, or both, washed plant and animal matter down to the lake bottom, and it ended up in the sediments. The problem was to find material that wasn't already hundreds of years old when it was deposited. Kelsey and his colleagues decided that the best approach would be to pick out things that would normally decay quickly if not buried in sediments—leaves of deciduous trees, leaf or flower buds, delicate insect wings—and to avoid fragments of wood, seeds, or anything else that was resistant to decomposition and might possibly be significantly older than the layer they wanted to date. That seems to have been a successful strategy, because their radiocarbon ages for the disturbance events are quite consistent from core to core throughout the lake.

Kelsey and his colleagues reported a total of sixty-one carbon-14 dates for the Bradley Lake sediments, which enabled them to construct an accurate time framework for great Pacific Northwest earthquakes over approximately the past 5,000 years (see figure 23). This is probably the most complete record we have, because it documents the occurrence of large tsunamis that sweep along the entire coast. Coastal submergence, on the other hand, is more likely to be restricted to regions close to the earthquake location.

Kelsey's results, like Atwater's, show that Cascadia zone earthquakes occur in clusters, with long gaps between them that sometimes last more than a thousand years. Kelsey and his colleagues concluded that the largest earthquakes (as inferred from the most severe disturbances of the Bradley Lake sediments) tend to occur at the beginning and end of the clusters (see figure 23). They also suggested that the 1700

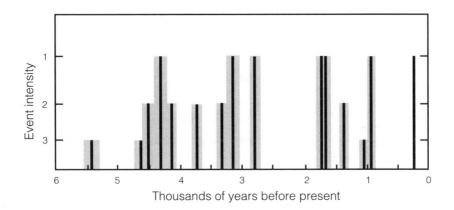

Figure 23. The record of large earthquakes along the Cascadia subduction zone, as determined from sediments in Bradley Lake, Oregon. Each dark vertical bar indicates an earthquake; its height is an approximate measure of intensity. The shaded region around each bar denotes the uncertainty in the radiocarbon date; the most recent event, with no uncertainty indicated, is the earthquake/tsunami of 1700 documented by Satake, Atwater, and others. Kelsey and his colleagues identified three intensity levels based on the Bradley Lake sediment records (labeled 1, 2, and 3 on the vertical axis): Category 1 earthquakes generated large tsunamis causing significant sand and saltwater incursion into Bradley Lake. Category 2 events generated tsunamis of lesser height and affected the lake less severely. Category 3 events resulted in ground shaking that caused disturbance of the sediments but generated no tsunami large enough to spill sand and salt water into the lake. Based on data in Kelsey et al., *Geological Society of America Bulletin* 117 (2005): 1009–32.

earthquake may be the beginning of a new cluster, because it follows a substantial gap in the record. It is possible that this is an overinterpretation of the data, but it is nevertheless evident that the current quiescence of the Cascadia subduction zone is no guide to its long-term behavior. Even if you take the view that the record is not long enough to discern a pattern, a simple average indicates that great earthquakes have shaken the region every 400 to 500 years. The last one took place just over 300 years ago. This information still isn't terribly useful if you are planning to buy a house in Seattle—unless, of course, you plan to live as long as Methuselah, in which case it would be a bad decision. But the record is not very reassuring for inhabitants of the Pacific Northwest, either.

Measuring very accurate ages on samples as small as the tiny twigs and insect wings from Bradley Lake sediments has been made possible by improvements in radiocarbon dating methods, some of which we will explore in chapter 9. Such analyses are a far cry from the crude first attempts by Bill Libby, Jim Arnold, and Ernie Anderson to determine the age of an Egyptian pharaoh's tomb. (Their work can only be called "crude" when judged by today's standards, however; at the time, it was cutting edge.) The ability to make very precise measurements has also made the role of the radiocarbon calibration curve all the more important for age determinations. What good is it to carry out highly accurate analyses if the calibration curve used to calculate the true age of a sample is uncertain?

We saw earlier that the calibration curve had its origins in the work of Hans Suess and others, who investigated the past constancy of radiocarbon in the atmosphere by measuring the carbon-14 content of wood already dated by tree-ring counting. Suess's data from the 1960s reached back about 7,000 years. By dint of tremendous effort by many different laboratories, literally thousands of tree-ring patterns have now been pieced together to extend the record to over 12,000 years. This is the backbone of the currently accepted calibration curve, because the tree-ring dates are absolutely accurate—the years can be counted off, one by one. There is hope that this record can be extended much further into the past by finding even older "fossil" trees preserved in marshes, bogs, and lakes, like the gigantic kauri trees from New Zealand that lived as long as 1,000 years. If continuously overlapping portions of the tree-ring patterns can be found, it may even be possible one day to calibrate the entire range of radiocarbon dating this way. But, for the moment, other approaches are necessary for older portions of the curve. Accurate calibration is extremely important for that earlier part of the carbon-14 dating range, because, for some time intervals, the difference between raw "radiocarbon ages" and the corrected calendar year ages can be hundreds, or sometimes even thousands, of years.

Because of the importance of accurate calibration, an international group of radiocarbon experts began meeting to ensure that the calibration was consistent across the spectrum of researchers involved in carbon-14 dating. Periodically they publish their best estimate of the calibration curve; the most recent appeared in the journal *Radiocarbon* in 2004. Beyond the tree-ring record, scientists have to rely on other types of analysis to provide independent dates. The most valuable materials for this task turn out to be from the oceans: corals, and the fossil shells of small planktonic organisms. Both can be dated by a method not discussed in this book, which is based on isotopes with short half-lives that are present in ocean water (these isotopes are part of the "uranium decay series" discussed in chapter 2). Corals and plankton incorporate the isotopes as they grow, much as all living organisms incorporate carbon-14; when they die, the radioactive clock starts ticking as the isotopes decay away. Although this technique has been refined and perfected in recent years, the dating uncertainties are still much greater than those for tree-ring counting, adding additional uncertainty to the calibration curve beyond the current tree-ring limit of 12,000 years.

Nonetheless, the existing calibration is an impressive achievement—large data sets including many hundreds of analyses have been scrutinized and screened, and the most reliable results selected to define the presently accepted curve. A glance at figure 9 on page 85, or figure 25 on page 214, both of which show portions of the calibration curve, gives an indication of just how consistent the calibration ages are from one data point to the next. Even though there are sharp "wiggles" in the curve, there is little or no scatter. One researcher commented that the calibration curve is not just a thin line on a graph; it is a ribbon—and progress is measured by the fact that, with each new version, the width of the ribbon decreases. This is just another way of saying that the uncertainty inherent in the age determinations is decreasing. Next time you read a newspaper article about radiocarbon dating of a new archaeological find, you can be pretty confident that the reported age is accurate.

We should, however, add one cautionary note. The "official" 2004 calibration published in *Radiocarbon* extended back only to 26,000 years before the present. By general agreement, essentially all published radiocarbon dates in this age range are based on this curve. But, for older ages, up to approximately 50,000 years—close to the limit for radiocarbon dating (samples much older than this contain too little carbon-14 to measure)—there are discrepancies between the calibration data sets measured by different research groups, and until the reasons for these discrepancies have been worked out, it is not clear which data sets should be used. The discrepancies have led to a certain amount of scientific feuding, because some researchers insist that only uncalibrated "radiocarbon ages" should be published for this time interval, while others—not willing to wait for consensus to emerge—have used one or another of the not yet officially sanctioned data sets, or some average of them, to extend the calibration curve and correct their radiocarbon age results. When agreement *is* reached on an official version of the curve, some dates arrived at in this way may have to be revised. Extending the tree-ring data will certainly help to solve the problem. In the meantime, dates above 26,000 years remain slightly more uncertain than those in the earlier part of the radiocarbon range.

One of the most important reasons so much attention is paid to the older parts of the calibration curve is that the period between about 20,000 and 50,000 years ago is a crucial one for the development of *Homo sapiens*. It encompasses the time of major migrations of our ancestors from Africa and the Middle East into Europe, Asia, and even Australia, and the development of the first known examples of art in the form of "jewelry" and cave painting in Europe. It is also a period in which there were large climate swings. Making sure the radiocarbon dates through this period are accurate is a prerequisite to unraveling the factors that influenced the dispersal and development of modern humans.

In Europe and in the Middle East, where many of the remains and artifacts of *Homo sapiens* habitation from the past 50,000 years have been found, the chronology of climate change is well known. Much of the information comes from ice cores in Greenland, which have provided detailed information about climate in the North Atlantic over the past 100,000 years. The ice has annual laminations reminiscent of tree rings, and the chemical properties of each layer are directly related to the temperature when the ice formed, which provides clues to the average climate. By counting back layer by layer—a tedious but quite precise process—the annual bands can be dated with unequivocal accuracy, often to within a few years, through the entire time span accessible to radiocarbon dating. To make maximum use of the ice core data in determining how climate affected the migration and development of *Homo sapiens* in Europe requires that the accuracy of radiocarbon dates for artifacts is as close as possible to that for the climate information.

In many cases, the most readily available samples for dating archaeological sites are pieces of bone, either human or animal. They are rich in carbon, and they obviously come from living beings. Provided these bone fragments are unambiguously part of the soil layer or archaeological horizon being dated, they would seem to be ideal candidates. However, because bones are porous and easily contaminated, the radiocarbon age determinations have proved problematic. As little as 1 percent contamination of a 30,000-year-old bone by "modern" carbon lowers its measured age by almost 3,000 years, and, for older samples, the effect is even greater. The major culprit seems to be groundwater that percolates through soil and rock, picking up along the way organic compounds that contain modern carbon. Bones that are continually bathed in groundwater sometimes incorporate these compounds. In one study, radiocarbon dates for bones retrieved from equivalent soil layers—and therefore presumably the same age—both from inside a cave (where the soil layers are relatively dry) and from outside (where the soil layers have been subjected to seeping groundwater) showed large differences in radiocarbon content, the bones from inside giving much older ages. The outside samples, apparently, were contaminated with modern carbon.

Because of the importance of bone for archaeological dating, scientists have made great efforts to solve the contamination problem. Fortunately, they have met with considerable success. Most laboratories now use a set of procedures designed to extract very pure samples of collagen—a fibrous protein that is the primary constituent of connective tissue—from their bone samples. This approach seems to remove nearly all contaminants, and, when only the purified collagen is used, the radiocarbon results are consistent with other available age information. Dates measured using this procedure, coupled with improvements in the data underlying the calibration curve, have led to changes in the accepted ages of several important European archaeology sites. In turn, the revised ages have forced researchers to reassess their understanding of migration patterns for *Homo sapiens* across Europe, and to revisit the question of interaction between modern humans and the Neanderthals, who had been the dominant hominid species in Europe for at least 100,000 years prior to the arrival of *Homo sapiens.*

Paul Mellars, an archaeologist at Cambridge University, England, is one of those who have examined the improved radiocarbon dates closely. In a 2006 paper in *Nature,* he proposed that modern humans spread through Europe much more quickly than had previously been thought, and that the period of overlap with their Neanderthal relations was considerably shorter than earlier radiocarbon results had indicated. Perhaps those don't sound like earthshaking findings, but both are crucial for formulating models of how we advanced as a species. The rapid population of Europe by *Homo sapiens* implied by the new results suggests that the early humans were better equipped to deal with adverse environmental conditions—specifically the cold of a glacial period— than had been believed by earlier researchers. And, if the overlap between modern humans and the Neanderthals really was short, the reason could be that the new invaders quickly outcompeted the earlier inhabitants, forcing them into extinction.

Mellars used the revised radiocarbon dates to trace out, in a rough manner, the routes taken by modern humans as they moved into Europe. From the Middle East, some 47,000 to 49,000 years ago, *Homo sapiens* moved north and west into Europe, following two distinct pathways, one along the Danube River, the other along the north shore of the Mediterranean (see figure 24). By 41,000 years ago, *Homo sapiens* was widely spread through Europe, colonizing parts of Italy, northern Spain, Germany, France, and the region north of the Black Sea. This dispersal occurred under a climate regime much harsher than exists in Europe today. Mellars concluded that the improved radiocarbon dates allow no more than 6,000 years of overlap (from about 41,000 to 35,000 years ago) between *Homo sapiens* and the Neanderthals in most of Europe, and perhaps as little as 2,000 years in some areas. Most earlier studies had estimated that the two groups coexisted for roughly twice as long.

Because of the uncertainties in the calibration curve beyond 26,000 years ago, not all of Mellars's archaeological colleagues agree with his conclusions. However, if he is right, the potential for interaction between modern humans and Neanderthals was more limited than once thought. Ever since evidence was unearthed that the two groups shared common territory in Europe at about the same time, both archaeologists and the general public have been fascinated with the possibility that they met and interacted. Neanderthals have often been caricatured as dull, plodding creatures, contrasting strongly with modern humans (although this is now thought to be an inaccurate and unfair comparison). But was there cultural exchange or even interbreeding before the Neanderthals died out? Or were the two groups simply fierce competitors? Neanderthal tools and implements tend to be less varied than those of roughly contemporaneous *Homo sapiens,* although archaeological sites from near the end of Neanderthal existence contain some tools that closely resemble those of modern humans. Was this imitation, or cultural transfer? DNA analyses have been used to search for genetic evidence of interbreeding between the two groups, but so far none has been found. If that result continues to be upheld, it supports the conventional

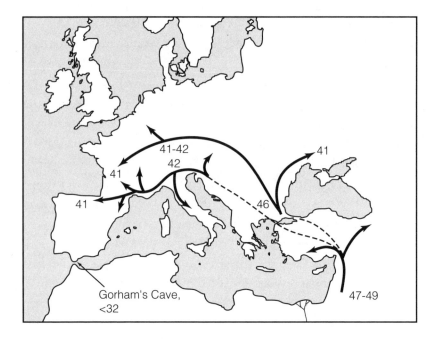

Figure 24. Migration times and routes for the entrance of modern humans into Europe based on revised radiocarbon dates and analysis by P. Mellars. Numbers show the revised dates for archaeological sites in thousands of years. The northern route is along the Danube River; the southern route is more or less coastal. Also shown is the location of Gorham's Cave, where Neanderthals apparently survived until at least 32,000 years ago. Based on data in P. Mellars, *Nature* 439 (2006): 931–35.

characterization of the Neanderthals as a separate species, in spite of the fact that we share a common ancestor. For many of the questions surrounding the Neanderthals and their coexistence with *Homo sapiens,* accurate dating is crucial.

The climate record from Greenland ice cores shows that, at about the time Mellars's data indicate that *Homo sapiens* dispersed rapidly across Europe, there was a brief amelioration of the frigid ice age climate, which may have been a factor in their expansion. But, by 35,000 years ago, near the time when most previous evidence suggested that the Neanderthals disappeared, Europe was descending toward the very coldest part of the last glacial interval. That *Homo sapiens* survived and the Neanderthals did not is often taken as an indication that the more modern hominids were better equipped to withstand the harsh ice age environment. But it may be that the Neanderthals didn't disappear quite so quickly. In 2006, just a few months after Mellars's work appeared in print in *Nature,* another study challenging his conclusions was published in the same journal. It reported new radiocarbon dates for Neanderthal habitation of a cave in Gibraltar, and claimed that these hominids survived much longer than had been suspected. If that is true, the overlap with *Homo sapiens* may have been considerably greater than Mellars thought.

Gorham's Cave is tucked away on the most southerly promontory of Spain. The new research indicates that it was inhabited by Neanderthals 32,000 years ago, and perhaps more recently, extending the known time of their existence by at least several thousand years. No human remains were found at the Gibraltar site, but tools typical of the Neanderthals were, strong evidence for their presence. The radiocarbon ages were measured on small fragments of charcoal from the same layers that contain the tools, and were pretreated to remove contamination. The large international team that reported the new dates is continuing work at Gorham's Cave, and, if their first results are corroborated, it will mean that at least one small band of Neanderthals were able to survive in their Mediterranean hideaway in spite of the weather and pressure from *Homo sapiens* invaders. Whether they were the last of their kind is a question that will probably never be answered. There may have been other small surviving groups living in isolation in southern Europe as well, not—or, at least, not yet—identified through any fossil evidence. But it is a poignant thought that the inhabitants of Gorham's Cave may have lived out their lives never knowing that theirs would be the last Neanderthal legs ever to climb Gibraltar's steep rock, and the last Neanderthal eyes to survey the blue Mediterranean.

Improvements in the radiocarbon calibration curve—especially reductions in the width of the ribbon of uncertainty—have been important at the younger end of the age range, too. A case in point is the precise dating of the great eruption of the Grecian volcano Thera, which occurred some 3,600 years ago. Archaeologists have long been interested in this event because it spread volcanic ash throughout the eastern Mediterranean, leaving a layer that provides an important marker horizon for Bronze Age archaeological sites. But the difficulty lay in dating the eruption. Techniques conventionally used for measuring the ages of volcanic ash, such as potassium-argon or uranium-lead dating, aren't applicable for such a recent event, or could only provide ages with very large uncertainties. And, on the face of it, volcanic ash can't be dated using radiocarbon either, because ash doesn't contain any living material.

The Thera volcano has a large central crater, the legacy of eruptions that occurred much earlier than the 3,600-year event. Only parts of Thera's rim poke above sea level today, forming the several islands of the popular Greek tourist destination of Santorini. The eruption that took place 3,600 years ago, and the earthquakes associated with it, generated huge tsunami waves that swept across the Aegean, damaging and destroying ships and settlements. Communities on the islands of Santorini that weren't submerged by the tsunami were completely buried in volcanic ash. One settlement, known as Akrotiri, is currently the site of extensive archaeological investigations—it has been compared to Pompeii, which was similarly engulfed in volcanic ash from the eruption of Mount Vesuvius in A.D. 79. But, in addition to burying villages, the eruption of Thera also buried living vegetation. In 2006, Walter Friedrich, of the University of Aarhus in Denmark, together with a group of colleagues, used a piece of that buried vegetation— together with the improved radiocarbon calibration curve—to obtain a very precise date for the eruption.

As the searingly hot ash fell, most vegetation ignited and burned to charcoal. However, Friedrich and his colleagues made a rare find: part of an olive tree that had been buried intact and remained unburned. Remnants of leaves and tiny twigs made it clear that the tree had been alive when it was engulfed in ash. Annual growth rings are notoriously difficult to discern in olive trees, but, by using X-rays, the researchers identified seventy-two rings in a portion of the tree branch that still had its outer bark intact. Here was a record spanning almost three-quarters of a century, extending right up to the instant of the eruption.

Even with modern techniques, samples taken from a single tree ring, or even several rings, are difficult to measure—especially if the entire sample is small, as was the case for the Santorini olive tree branch. Friedrich and his colleagues cut out four separate samples for analysis, the largest comprising wood that had grown over a period of twenty-four years, based on the rings. Each was measured for carbon-14 and its "radiocarbon age" calculated. But, because the exact number of calendar years between each pair of samples was known from tree rings, it was possible to "wiggle match" the results with the calibration curve in a way not possible with a single sample. Essentially, the data points, with fixed radiocarbon ages, and also fixed with respect to one another through tree-ring counts, could be shifted horizontally as a group of four until the best fit to the calibration curve was found (see figure 25). Without a very detailed calibration for the time interval around 3,600 years ago, such an approach would be impossible.

The radiocarbon results showed that the very last annual ring of the olive tree grew between 1605 and 1621 B.C., placing the Santorini eruption within this very narrow time interval. A second high-precision carbon-14 date on seeds from a storage container found in the buried village of Akrotiri overlaps the olive branch age, confirming its accuracy. These results have implications far beyond Santorini itself, because archaeologists working in the region—through analysis of cultural connections between Egypt and the Aegean region—had generally placed the eruption around 1520 B.C., or even later. The new date shows it occurred a century earlier, which will require significant revisions in all chronology that used the widespread ash layer as a reference horizon.

Estimates of the size of the Santorini eruption put it among the very largest of the past several thousand years. It had global consequences—

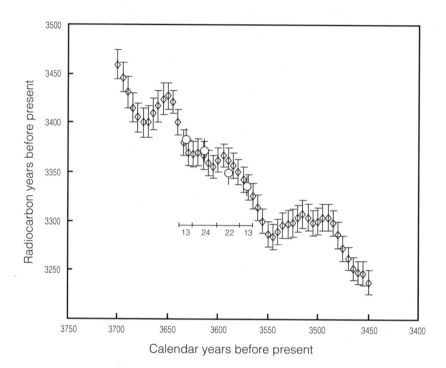

Figure 25. Four radiocarbon ages (open circles) for portions of the Santorini olive branch matched to the wiggles of the radiocarbon calibration curve (the latter shown as diamond symbols with associated vertical bars denoting uncertainty). Fitting the data in this way was possible because tree-ring counting defined the exact number of years between each data point. The total time span represented by the tree branch (72 years), and the number of rings included for each analysis, are shown by the line below the calibration curve. The last tree ring falls at the right-hand end of this line, and dates the year of the eruption. Based on Santorini data in W. L. Friedrich et al., *Science* 312 (2006): 548, and calibration curve data from Paula J. Reimer et al., *Radiocarbon* 46 (2004): 1029–58.

ash from the volcano has been found in ice cores from Greenland, and some researchers have suggested that crop failures in China around that time, as well as cold damage to trees in Ireland and California, could have been caused by the eruption. Clearly, such links can be made only if the dating is accurate. And, for that reason, as you might imagine, when work like that on the Santorini olive branch gets published, it makes waves in the archaeological community. It may even make it into the science pages of a few newspapers. But the radiocarbon dating study that generated by far the most intense interest worldwide is one that in many ways is much less important—at least in terms of its scientific implications. It focused on an artifact that is now probably the most intensely studied man-made object that ever existed, and which is certainly the most famous ever dated using radiocarbon.

The Shroud of Turin is a large piece of linen cloth that bears the indistinct, full-length image of a man—bearded, and apparently crucified. For centuries it has been (controversially) claimed to be the burial cloth of Christ. However, its history is—if you'll excuse the expression—shrouded in mystery. The first record of its existence is from the 1350s, when it was given to a church in Lirey, in the east of France, by the widow of a local knight. How the knight obtained the cloth in the first place is not known, but within years of its appearance in the church it was attracting pilgrims from near and far. They came to see what they believed was a holy relic.

From the 1300s to the present, the shroud has had a colorful history, surviving at least one fire (and getting scorched in the process), being condemned as a fake, sold and moved to another part of France, and eventually ending up in the cathedral in Turin, Italy. It is quite possible it might have remained there, hidden in obscurity, had not an Italian photographer decided to take a picture of it in 1898. Much to his surprise, when he processed the film in his darkroom, the indistinct markings on the cloth were much sharper on his photographic negatives. The vaguely defined face on the shroud image stood out as clearly as if it were a positive photograph. This has led to the idea that the image itself is actually a kind of "negative" (two negatives make a positive). Not surprisingly, the 1898 photographs created quite a sensation; it was around the time of Roentgen's discovery of X-rays, and there was great interest in how the image might have been formed. From that time until today, the shroud has been the focus of a huge amount of research—and speculation. Theories about the origin of the image are numerous. They range from the idea that it was somehow formed by supernatural "radiation" emanating from the body of Christ, to the possibility that it is a hoax produced by Leonardo da Vinci experimenting with a (very early) kind of photography, to its being simply an image painted on the cloth by an unknown artist.

One way to test the veracity of the shroud story would be to determine its age accurately. Radiocarbon dating is an obvious approach; the cloth was woven from once-living plant material, most likely harvested no more than a few years before manufacture of the linen, and the probable age is well within the range of the method. Dating the Shroud of Turin was an aspiration of radiocarbon researchers for many years, but the difficulty was that dating required a significant amount of cloth to be cut from the shroud—and its keepers in Turin were loath to allow that. However, with the advent of the new technique of accelerator mass spectrometry (discussed at greater length in chapter 9), sample size requirements were greatly reduced, and the experiment became feasible. So, in 1983, the British Museum coordinated a comparison study to determine whether it would be possible to accurately date small samples of cloth. The experiment involved six different radiocarbon laboratories—four of them using the new method and two using more conventional techniques. The results were promising. Each laboratory had been sent three small textile samples for dating, the ages of which were known to the British Museum, but not to the analysts. The dates they sent back were accurate and consistent with one another—and all the work had been done on samples of a size that would probably be acceptable to the authorities in Turin. The stage was set for measuring the real thing.

Even so, getting a sample of the Shroud of Turin for analysis was a very different exercise from digging an olive branch out of the volcanic ash on Santorini, or from scraping a piece of charcoal from the soil of a Gibraltar cave. Final approval had to come from the Vatican, the official owner of the shroud. In 1987, the papal authorities finally selected three of the six laboratories from the original comparison study and agreed to provide them with samples. The laboratories were internationally located—in Tucson, Arizona; Oxford, England; and Zurich, Switzerland—and all were centers for the accelerator mass spectrometry technique. The British Museum was asked to help authenticate and document the samples, and to examine the analysis results. The actual sampling took place in April 1988, in the sacristy of the Turin Cathedral, and was a solemn affair attended by the archbishop of Turin, textile experts, and representatives of the British Museum and the laboratories involved. A small strip was cut from the shroud, divided into three, and each piece sealed in a stainless steel container for transport to the laboratories. Similar amounts of three different "control" samples, each of approximately known age, were also distributed. The labs were not told which sample was which.

The results of this elaborate exercise were published in the February 16, 1989, issue of *Nature*. On the cover was a picture of the shroud, showing the face of the man reputed to be Jesus Christ. The paper describes the meticulous care each laboratory took with these samples—splitting them into several pieces for multiple analyses, and using various preanalysis cleaning procedures to ensure there was no contamination with modern carbon. Just as had happened in the earlier comparison experiment, the dates obtained by the three laboratories were consistent with one another. Using the most recent calibration data, the results indicate that the flax used to make the shroud was harvested sometime between A.D 1260 and 1320. For technical reasons—there is a large wiggle near 1300 in the calibration curve—it is possible that the permissible age range extends to almost 1390. But it is most likely that the true age is within the earlier part of the range.

It is clear from the radiocarbon dates that the shroud is not 2,000 years old—or at least the analyzed sample is not 2,000 years old. The fact that the measured age falls close to the time of the first recorded appearance of the shroud suggests that it indeed originated near the end of the thirteenth century, and, even if the image is not a deliberate fake, it is certainly not an original image of Jesus Christ. Still, there are some who want it to be authentic and who have proposed various scenarios that would account for the observed carbon-14 content (which almost everyone accepts as being correct) and yet still allow the possibility that the shroud is really 2,000 years old. Some of these are, to say the least, very inventive. But the doubters are encouraged in their hopes by such details as the fact that the weave in the shroud—a herringbone pattern—is consistent with its having been made at the time of Christ, and also by the discovery of pollen grains in the cloth that some experts have declared come from plants found only in the vicinity of Jerusalem.

If permitted by the Vatican, the debate about the shroud's origins could eventually be settled by further radiocarbon analyses. A few laboratories now have the capability to analyze such small samples that it might even be possible to measure the carbon-14 content of a collection of the putative Middle Eastern pollen grains. But even if these, too, turned out to be much younger than 2,000 years, it is unlikely that everyone would be convinced. Interest in books like *The Da Vinci Code* shows that there is such a fascination with religious relics that even the best science may not be able to sway peoples' convictions. However, at the very least, this work brought radiocarbon dating to the attention of many people who otherwise might be entirely unaware of its potential. Try typing "Shroud of Turin" into Google. You'll be led off in so many directions that you could spend the next week navigating through the volumes of material that exist about this simple piece of linen. Nearly all those discussions have substantial sections dealing with radiocarbon dating. Just remember not to believe everything you read.

CHAPTER NINE

More and More from Less and Less

This chapter is about modern dating practices, which does not mean meeting someone over the Internet. Rather, I'd like to take a brief look at a few of the advances of recent years that make the field of geochronology so pervasive in the earth sciences today, and so exciting. The inventiveness and power of some of the methods used to measure the past, I believe, are not given enough credit. Reports of radiometric age determinations are close to being commonplace; the more "exciting" ones are almost as likely to pop up in the newspaper or on TV as in a scientific journal, and it might be tempting to think that after the discovery of radiocarbon dating, the determination of the Earth's age, and the development of potassium-argon dating, geochronology became routine, more a technical exercise than a creative one. The most common of the naturally occurring radioactive isotopes had been exploited. What was left to do? That sentiment reminds me of a recent *New Yorker* cartoon in which two men dressed in animal hides sit on a rock. "Og invented fire," says one, "and Thorak invented the wheel. There's nothing left for us." But it is worth remembering that many commentators, including quite a few physicists, said essentially the same thing about physics not long before the discovery of radioactivity, relativity, and quantum mechanics proved them completely wrong. While the recent advances in geochronology may not be of quite the same order as those discoveries, they do show that the field is just as vibrant.

One of my colleagues, a brilliant scientist, continually challenged everyone around him to seek new discoveries in geochronology. We sometimes jointly taught a course in isotope geology, and, to the extent schedules permitted, each of us would attend the other's lectures, so I got to see his teaching approach firsthand. He was perhaps a bit naïve about the likelihood that students would complete assignments that were entirely optional and that they knew would not be graded—he ended most of his classes by giving them just such a challenge. Usually it was a problem directly related to things he had talked about during the class, but sometimes it was more general, perhaps related to scientific principles encountered in everyday life. On several occasions over the years, he asked students to develop a new dating method based on radioactive decay. He was quite serious (perhaps because he had developed new methods himself). He would hand out a copy of the periodic table and explain that, among the ninety elements that occur on Earth, there were some radioactive isotopes not yet being used for geochronology. As it turned out, none of our students ever came up with a suggestion that led to development of a wholly new technique. But some did contribute imaginative ideas, and, probably more important, my colleague's challenges jolted all of them into realizing that there really are still things out there to be discovered.

Far from worrying about possible stagnation in their field, geochronologists like my colleague are always looking for new ways to tell time. Sometimes they have done so with one of the naturally occurring radioactive isotopes that had never previously been used for dating. Using these isotopes for radiometric dating might now be possible, for example, through the development of new, more sensitive instrumentation. Or sometimes a researcher will devise a new twist for a tried-and-true technique, making it possible to analyze materials that could not be dated before. However, more than a century after the discovery of radioactivity, the "easy" dating applications are well established, and breakthroughs are rare—which is why it was difficult for our students to devise an entirely new method.

Many of the most important advances in the field have been made through the cumulative effects of small improvements in scientific procedures and instruments. The push to develop methods that are simultaneously more accurate and also capable of analyzing smaller and smaller samples has been particularly important. In the case of precious samples from the moon, or a rare meteorite, or a valuable archaeological artifact, finding ways to measure very small samples has been a necessity. But, quite often, advances in microanalysis—for example, development of the capability to date a single grain in a rock—have also opened up a whole new range of questions for investigation.

Although analytical instruments are to some degree just a means to an end, honing them to their current level of performance has required very close collaboration between instrument makers and instrument users. For a long time, there was no distinction between the two, because the scientists interested in measuring the ages of things had to design and build their own instruments. Now almost none do. However, even off-the-shelf instruments purchased from a commercial manufacturer are usually tailored to the needs of the laboratories that order them, and require much back-and-forth discussion during manufacturing and testing. This is not the place to discuss in detail the technical advances that have made age determinations so reliable and precise, but it is worthwhile enumerating some of the goals that led to those improvements. In general terms, several themes have been important as geochronologists designed and updated their instruments: (1) improving the precision with which measurements can be made, thereby reducing the uncertainty in age determinations; (2) making it possible to measure much smaller samples; (3) developing microanalysis techniques for analyzing samples in situ; and (4) speeding up the analysis process so more samples can be analyzed in the same amount of time.

Progress toward many of these goals got a jump start in the 1960s as laboratories—especially in the United States—geared up for analysis of samples returned from the moon by the Apollo program. Everyone realized there would be a premium on making accurate measurements of this rare material as quickly as possible. People were eager to know what history the moon rocks held—and, because it wouldn't be possible to go back to the rock outcrop and take another sample next field season, it was obvious there would never be much material to work with. Precise analysis of small samples was therefore crucial.

What kinds of instruments are used today in dating studies? Broadly speaking, they take one of two general forms, both of which we have already encountered in previous chapters: counters—instruments that measure the number of radioactive decays that occur in a sample—and mass spectrometers, which measure the quantity of specific isotopes in a sample. Both have their origins in the early part of the twentieth century, shortly after Marie Curie discovered radioactivity and Ernest Rutherford illuminated the structure of atoms. By a wide margin, mass spectrometers are the most commonly used instruments for modern geochronology; counters, for reasons that will become clear below, are employed less frequently, although for some applications they may be the only choice.

The first true mass spectrometers were built just after the First World War. The Cavendish Laboratory at Cambridge University—where Ernest Rutherford worked after leaving New Zealand—was a hotbed of physics research, and predictably became a center for mass spectrometry. Francis Aston, working there with his (and Rutherford's) mentor J. J. Thomson, used this instrument to show beyond any doubt that most of the chemical elements in the periodic table are made up of several different isotopes (he discovered no fewer than 212 of the naturally occurring isotopes and was awarded the 1922 Nobel Prize in Chemistry for his work; like Rutherford, the prize transformed him from a physicist to a chemist!). Aston is often credited with being the inventor of the mass spectrometer in 1919, but Arthur Dempster, a physicist at the University of Chicago, actually built a mass spectrometer in 1918 that is closer in design to most modern instruments. However, Dempster did not pursue the extensive survey of isotopes that Aston did.

With Aston's discovery, it became quite clear that age measurements based on radioactivity would require the measurement of isotopes, not just bulk analysis of the radioactive parent element and its daughter, and this would require a mass spectrometer. It took many years to develop ones that could make accurate isotope measurements on complex materials like rock samples, but, ever since, mass spectrometers have been the workhorses of geochronology.

How do these ingenious instruments work? Even the most sophisticated modern ones are conceptually quite simple, with just a few important parts. They are designed to sort out atoms or molecules on the basis of their mass, and different types of mass spectrometers do so in different ways. A common approach, the one employed for the very first mass spectrometers, is to use a magnetic field. In a typical arrangement, the sample to be analyzed is ionized—that is, its atoms are given an electric charge—and the ions are fired at high speed through a carefully controlled magnetic field. The field causes their paths to curve—a little for an ion of high mass, and a lot for one that is not so heavy. By adjusting the magnetic field, researchers can direct ions of a particular mass into a fixed "collector"—a device that measures their abundance. The whole of the mass spectrometer is kept under a high vacuum so the ions speeding through it have an unimpeded journey and don't collide with gas molecules along the way. In some instruments, there is an array of collectors so ions of several different masses, which follow paths with slightly different curvatures, can be detected simultaneously. When measuring a sample of pure lead, for example, these collectors might be positioned so that all four of the lead isotopes (with atomic masses of 204, 206, 207, and 208) are measured simultaneously and their relative abundances recorded. Figure 26 shows a mass spectrometer of the type used for uranium-lead dating.

Mass spectrometers are incredibly versatile, and are used for a range of purposes far beyond geochronology. They can detect steroids in an athlete's urine and determine whether a sample of uranium is natural or has been processed in a nuclear weapons program. Several miniaturized

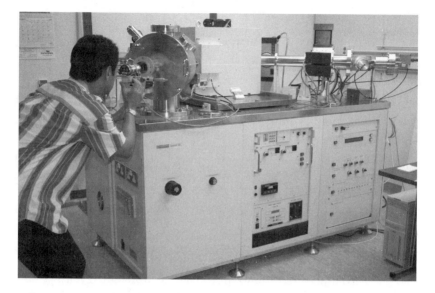

Figure 26. A mass spectrometer at the Scripps Institution of Oceanography of the University of California at San Diego. Samples are placed in the chamber into which geologist Pat Castillo is peering. Ions produced by heating are sent along a stainless steel tube through the magnet (large gray block, top center), where they are deflected along a curved path into collectors at the top right of the picture. Photo courtesy of Pat Castillo.

versions have been sent into space; in 2005, the probe that landed on Titan, one of Saturn's moons, carried a mass spectrometer that sent back data about the composition of the moon's atmosphere and surface.

The great advantage mass spectrometers have over counters is that, in principle, they can detect and record every atom in the sample. In contrast, a counter records a "count" only when a radioactive atom decays—by detecting the particle that is emitted, not the radioactive atom itself. The advantage increases as the half-life of the radioactive isotope increases. Think about a hypothetical radioactive sample of 1,000 atoms. In principle, a mass spectrometer could detect all these atoms during a single, short measurement. Using a counter, however, and assuming our hypothetical isotope had a short half-life—say, one day—just half the sample (500 atoms) would decay and be recorded over a twenty-four-hour period. During the next twenty-four hours, half the remaining atoms would decay, and the counter would record an additional 250 counts. If the half-life were 10,000 years, however, you could forget about using a counter altogether. You would be lucky to record even a single decay in 10 years of measurement. That's far too long for even the most patient of graduate students.

The 5,730-year half-life of carbon-14 is short enough that counters are appropriate for its measurement, provided the sample is big enough. The technique was founded on this technology; Libby and his colleagues made their first measurements using counters, and, for several decades, it was the only method available because, as will be explained below, conventional mass spectrometers cannot measure radiocarbon. However, particularly if available samples are small, mass spectrometry is the analytical method of choice for radiocarbon dating today.

The specialized mass spectrometers used for carbon-14 measurements are very different from the one shown in figure 26. Initially, at least, they were gigantic versions that required warehouse-sized laboratories to house them. (More recently, in the past decade or so, smaller instruments have been designed.) Regardless of size, however, their development has been the single most important advance in the field of radiocarbon dating since Libby invented the technique. The important feature of these mass spectrometers, and the reason they are so large, is that they incorporate accelerators—devices that take ions from the sample and speed them up to tremendously high velocities before various processes are used to sort them out by mass. As a consequence, the method is generally referred to as accelerator mass spectrometry—AMS for short.

Why is such complicated (and expensive) instrumentation necessary? There are many technical details that bear on this question, but the simple answer is that conventional mass spectrometers can't discriminate between carbon-14 and other ions, such as nitrogen-14, that are very close in mass. Nitrogen is the major constituent of the atmosphere, and is virtually impossible to exclude during analysis. And a mass spectrometer doesn't care whether an atom is radioactive or not; it discrim-

inates solely by mass. Only by accelerating the ions to very high velocities is it possible to strip out the billions and trillions of interfering ions
and get a true carbon-14 signal.

The AMS technique permits very accurate measurement of a small
number of carbon-14 atoms in the presence of huge numbers of others,
which is why it is so effective for small samples. Whereas counters may
require several grams of carbon per sample, AMS analyses can be done
on a few ten-thousandths of a gram, and sometimes even less. This
makes it feasible to analyze many things that could never be measured
before, such as single seeds, microscopic fossils from deep-sea sediments, individual tree rings, or a few specks of charcoal from a Pale-
olithic cave drawing.

AMS is used to measure many isotopes in addition to carbon-14, but
by far its most diverse applications involve radiocarbon. In part this has
to do with the ubiquity and importance of carbon as a chemical element.
Not all AMS applications involving radiocarbon fall into the category of
"dating" problems—some simply take advantage of the fact that
carbon-14 occurs only in matter that has "recently" been part of a living
organism. In some contexts—for example, in studies of air pollution—
its presence or absence can serve as a kind of tracer of the source of the
carbon. The very high sensitivity of AMS makes it possible to separate
and analyze extremely small amounts of different constituents from a
sample of "polluted" air, for example, and to distinguish between compounds that originate from living things such as trees and animals,
which contain modern carbon-14 levels, and those that stem from
petroleum-based industrial products or fossil fuel burning.

The small-sample capabilities of AMS have also allowed researchers
to take advantage of an unanticipated outgrowth of the nuclear age. Although most scientists abhor nuclear weapons, they are not about to look
a gift horse in the mouth. When atmospheric testing of atomic bombs
began in the 1950s, the bombs, like the cosmic rays, produced carbon-14
in the atmosphere. But they did so in much larger amounts, resulting in
a huge spike, or "pulse," above the natural background level of carbon-

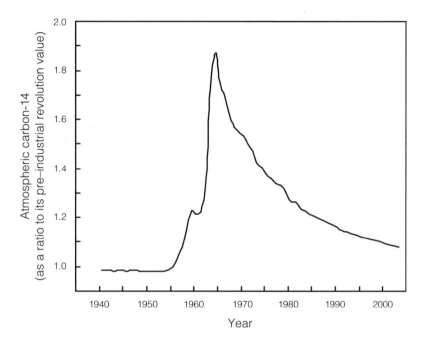

Figure 27. The pulse in atmospheric carbon-14 content due to testing of atomic bombs in the atmosphere. Units on the vertical scale show the concentration relative to that before the industrial revolution, expressed in the usual units of carbon-14 per gram of carbon. Note that before testing began, the atmospheric value was slightly less than 1.0 on this scale, owing to the dilution of carbon-14 by extra carbon dioxide (containing no carbon-14) introduced into the atmosphere by fossil fuel burning.

14 (see figure 27). This is one of the reasons that the carbon-14 content of modern carbon is referenced to 1950; since the atomic tests began, all living things have incorporated some bomb-derived carbon-14. The buildup of excess radiocarbon introduced by nuclear testing was rapid between 1955 and 1963. Then, with the cessation of atmospheric explosions, it began to decrease again—mostly because of its uptake and storage in living material, and its dissolution in the oceans. (Because of the 5,730-year half-life of carbon-14, radioactive decay accounts for very little of this change.)

This well-documented increase and subsequent decline of atmospheric carbon-14 is the basis for an entirely new type of radiocarbon dating. It is useful only over the past half century or so, and you might think there is no need for independent time determinations over that period. But, in fact, it turns out to be very useful. This nascent field has already seen some very ingenious applications. One of the earliest and most important—begun even before AMS analyses became commonplace—was the measurement of carbon-14 in the oceans. Radiocarbon from the atmosphere can enter the oceans only at the sea surface, so the nuclear testing spike provides an ideal tracer of how quickly that happens, and of how rapidly the carbon is mixed into and transported through the oceans. Like the radiocarbon produced by cosmic rays, the carbon-14 from atmospheric nuclear testing was quickly oxidized to carbon dioxide, so in reality these measurements trace the fate of CO_2 that enters the oceans from the atmosphere. This is especially important information, because it helps climate change researchers to understand how much of this greenhouse gas will be soaked up by the oceans as we burn more fossil fuels, and also how quickly that will happen. Since the 1970s, literally thousands of radiocarbon measurements have been carried out on samples from all parts of the world's oceans in pursuit of such knowledge.

Forensic scientists were not far behind oceanographers in exploiting the bomb-produced radiocarbon pulse. Because every part of the human body containing carbon is labeled with the atmosphere's carbon-14 signature at the time it grows, matching radiocarbon contents with the well-determined curve in figure 27 can provide dates for body parts. That sounds a bit gruesome, but such information can be of crucial importance in criminal cases or war crimes investigations.

Tooth enamel has turned out to be especially useful for such work, because it forms at specific times during a person's life and thus can be used as a very precise time marker. Numerous studies have shown that wisdom tooth enamel is the very last to grow, and that it is typically formed at age twelve, with very little variation. This means that any person on Earth who turned twelve after the first atmospheric bomb test—that is, anyone born after 1943—will have bomb-produced radiocarbon in their wisdom tooth enamel; those born earlier will not. The exact amount is an accurate indicator of the year the enamel grew, which can be read directly from graphs like figure 27. You can also see that there is potential ambiguity because of the pulselike nature of the graph; for most carbon-14 values, there are two possible years of enamel growth. However, there is a clever way around this difficulty. Enamel from other types of teeth—ones that grew before the wisdom teeth—can also be measured. This makes it possible to figure out whether carbon-14 was rising or falling during tooth formation, and thus to determine the year of wisdom tooth growth precisely.

One of the most definitive studies of this sort was conducted in 2005 by researchers from the Karolinska Institute in Sweden and the Lawrence Livermore Laboratory in California, who used AMS to analyze tooth enamel from twenty-two individuals with birth dates between the 1950s and the 1990s. When researchers added twelve years to the subjects' wisdom tooth ages, the results matched the known birth dates with an average variation of only about one and one-half years. This is far more precise than other methods available in forensic science, which typically produce age estimates that are valid only within five or ten years.

The diversity of radiocarbon dating and radiocarbon "tracing" applications depends to a large extent on the fact that carbon is integral to life. None of the other isotopes used in geochronology can match that special property, but this has not hindered their development. Each dating method has its own unique attributes and capabilities and has grown from a simple tool for age determination to an approach that can provide detailed information about complex geological processes.

When Harrison Brown gave Clair Patterson and George Tilton the task of using uranium-lead dating to measure the ages of Precambrian granites in the 1950s, they had to combine many zircon crystals for each analysis. Since then, improvements in conventional mass spectrometry and sample preparation procedures have made it possible, in the best cases, to make these measurements using just a few grains, or sometimes even a single zircon crystal. Still, large quantities of rock are usually crushed and processed to obtain a pool of crystals from which those for analysis are ultimately selected. Technicians may sit for days or weeks, peering through microscopes and sorting out the grains one by one; with experience, they learn to recognize which crystals are most pristine and most likely to give reliable results. But that is just the beginning. The selected zircons must be thoroughly cleaned to avoid lead contamination, usually by treatment with strong acids or by stripping away their outer portions through abrasion. The crystals are then dissolved in acid, and the lead and uranium they contain separated out by chemical means. Finally, those purified elements are loaded into a mass spectrometer for isotope analysis.

In the 1980s, however, Bill Compston and his group at the Australian National University developed an instrument that eliminated some of these steps, as mentioned briefly in chapter 5. It was an ion microprobe, a microanalytical instrument that is used in several different fields, but theirs was specifically designed for uranium-lead dating of zircons. Its most important attribute is that the sample is introduced into the instrument not as a purified element, but as a whole grain—or even as a slab of rock. In most cases, individual zircon crystals are separated from their host rock, embedded in epoxy, and polished flat, so that a cross-section of each grain is exposed. With the aid of a microscope, a thin beam of ions is focused onto a single small spot on the sample, and the bombarding ions blast away its surface, layer by layer. The atoms released from the crystal in this way are swept into a mass spectrometer for isotope analysis. During the course of a measurement, the ion beam drills a tiny hole into the crystal, typically less than a thousandth of an inch across (see figure 14 on page 125). Only a minute amount of the sample is used up in this process, and it is usually possible to make multiple analyses on a single grain.

Being Australian, Compston and his colleagues nicknamed their instrument SHRIMP. It is not one you can put on the barbie, though;

the acronym stands for sensitive high-mass resolution ion microprobe. Not to be outdone, the developers of software programs for analyzing data from instruments like the SHRIMP have christened them with names like SQUID and PRAWN. Who said scientists don't have a sense of humor?

The microscale analytical capabilities of SHRIMP and similar instruments have revolutionized the science of dating zircons. Although, for technical reasons, conventional mass spectrometry may occasionally be preferable, SHRIMP has been invaluable for highlighting the complexities that can exist within a single zircon crystal but remain invisible to conventional analysis. In part, the complicated nature of this mineral stems from its resiliency. Zircon crystals can survive severe metamorphism when other minerals don't, but, in the process, the surviving grains are sometimes overgrown by younger layers, and some of the lead generated by radioactive decay may diffuse into other parts of the grain—or leak out of it entirely. In the most ancient parts of the Earth's crust, rocks have experienced many episodes of metamorphism over their long lifetimes, and the zircon crystals they contain can be very complex, at least in terms of their uranium-lead characteristics. Although they remain single grains in a physical sense, and can be separated from a rock and handled individually, on a microscopic scale they may consist of multiple "domains" with different chemical characteristics and "ages." In such situations, it is difficult to interpret a conventional uranium-lead date even when it is measured on a single crystal. SHRIMP, however, is capable of measuring each area of a crystal and each generation of overgrowth separately, and in the best of cases can provide a complete chronology from the time of formation of the original rocks, through several metamorphic episodes, to the present.

Miniaturization and other novel approaches to sample analysis have wrought large changes in other radiogenic dating methods as well. In conventional potassium-argon analyses, for example, each sample is heated in an oven until it melts. Argon and all the other gases it contains (even solid rock contains dissolved water and gases such as carbon dioxide) bubble and diffuse out of the molten sample and are collected and purified. Pure argon is then separated and introduced into a mass spectrometer for analysis. In modern instruments, the entire procedure has been scaled down so that, compared with earlier times, quite small samples can be analyzed. But the ultimate accomplishment would be to measure single grains. In the early 1980s, Derek York, of the University of Toronto, came up with a way to do just that: heat the sample with a microlaser beam.

York's laser heating technique did for potassium-argon dating what SHRIMP has done for uranium-lead age measurements. In York's instrument, a narrow laser beam, smaller than a single mineral grain, is focused onto a sample in much the same way as an ion beam is in the SHRIMP machine. However, the strength of the laser beam can be varied continuously, just as lighting in a room can be controlled using a dimmer switch. This means the sample grain can be heated in steps by gradually cranking up the intensity of the laser, each step attaining a higher temperature than the previous one. At each temperature step, the argon is collected and analyzed, and an age calculated. In addition to providing a new capability for small-sample analysis—the tiny glass spherules from the K-T boundary sediments and the mineral grains used to date Lucy (see chapter 7) were both analyzed in this way—the laser probe can help resolve complexities by identifying small domains within individual grains that release argon differently at different temperature steps. In this respect, it has similarities to SHRIMP, because it can provide a chronology for crystals with a complicated metamorphic history.

New instruments such as AMS, SHRIMP, and the laser probe have been major factors in advancing the science of geochronology, but they have not been the only ones. Especially for the time since fossils became abundant in sedimentary rocks at the beginning of the Cambrian period, an important goal for geologists has been to increase the time resolution with which evolutionary processes and geological events can be examined. But all the radiometric dating methods, regardless of the type of instrument used, have inherent limits beyond which the uncertainties of individual ages—the "plus or minus" part of the result—can't be reduced. These limits have to do both with the measurement techniques themselves and with the uncertainties in the half-lives used in the age calculations, putting constraints on the ability of radiometric dating to resolve closely spaced events. Especially for investigating some evolutionary processes, paleontologists need better resolution than these constraints allow.

Currently, the absolute minimum uncertainties for radiometric dating are about 0.1 percent, although they vary depending on the nature of the samples and the procedures used in the laboratory. But even a minimum uncertainty of 0.1 percent translates to ± 400,000 years for a 400-million-year old sample, and this is very much a best-case scenario. The goal of many paleontologists is a resolution closer to 50,000 years. Only at that level can they address many questions having to do with the rates of evolution and extinction.

But is it even possible to achieve better time resolution than is attainable through direct measurements of individual samples? Perhaps surprisingly, the answer is yes. That accomplishment has involved the synergistic efforts of paleontologists, geologists, physicists, and mathematicians. Together they have amassed large amounts of geological data that are in some way related to time—from the appearance and disappearance of fossil species, to changes in ocean chemistry that are reflected in the composition of sedimentary rocks, and, of course, actual radiometric dates. Once collected, such data can be "sequenced," or put in a time order, with the aid of appropriate computer programs. If enough radiometric dates are available through the time span being examined, very high time resolution can be attained.

There is no single place on Earth where sedimentary rocks provide an uninterrupted record of the past 500 million years of Earth history, so the sequencing approach must rely on data from many different localities. Much of this information comes from field geologists who focus their work in a specific geographical region, building up a detailed and comprehensive picture of local geological history based on fossils, rock types, and virtually any other property of the rocks that can be measured. Locally, all these observations can be placed in a relative time sequence based on where they occur in a sequence of rock layers. Usually the data include some radiometric dates as well, which anchor the relative scale. The beauty of the new computer-based sequencing approach to correlation is that it can—in the words of Peter Sadler of the University of California at Riverside—merge such local records from different sites around the world to create a "global calendar of past events."

The geological literature is vast, containing literally thousands of scientific papers that describe the comings and goings of fossil species, the chemical properties of sedimentary rocks, and the radiometric dates for volcanic ash layers. An equal or perhaps even larger amount of data is proprietary, carefully guarded by the companies that extract oil and gas from sedimentary rocks, for whom time scales and correlation are very important. Only a small portion of this store of data has so far been examined using the new, computer-based sequencing approach to stratigraphy and correlation. But already the results are striking. In several studies, time resolution of around 50,000 years has been achieved by optimizing the sequence of thousands of sedimentary rock features over time spans of tens of millions of years—even with relatively few radiometric dates available through the time interval. The approach is not entirely problem-free— for example, a sequencing program may find more than one equally probable solution. But once such difficulties are identified, they can usually be resolved by locating additional datable sedimentary horizons and determining their ages. The importance of including as many radiometric dates as possible has prompted geologists to revisit sedimentary rock sections and look again for volcanic ash layers, which, because they are fixed time markers and are often widespread in their distribution, are especially valuable for sequencing studies. And the old saw that says you only find what you are looking for has proved true. Many "new" ash layers have been found where none had been reported before. Usually they had been overlooked because they were thin and inconspicuous, the products of small or distant eruptions. But the important thing is that most of them contain minerals that can be dated.

In some parts of the geological record, it is also possible to achieve high time resolution—but not actual ages—using an approach that does not involve radiometric dating at all: by using the Earth's orbital cycles around the sun to derive timing information. This may seem strange at first. What do orbital cycles have to do with the ages of rocks? But, in many places around the world—for example, at various localities in southern Europe—there are great stacks of ancient sedimentary rock layers that exhibit extremely regular and systematically recurring cycles of rock types. These have long attracted the attention of geologists, and the most spectacular examples are obvious even to a casual observer, the repeated pattern of layers almost as distinct as a zebra's stripes. The processes responsible for these features became known only when studies of deep-sea sediment cores showed that similar sequences were laid down on the ocean floor in the very recent (i.e., hundreds of thousands of years) past, the changing sediment types recording the ocean's response to climate changes induced by variations in the Earth's orbit. They are now known as Milankovitch cycles after the Serbian scientist Milutin Milankovitch, who in the first few decades of the twentieth century worked out a general mathematical theory relating the Earth's climate to its orbital parameters.

Milankovitch found that several aspects of the Earth's movements in space—the tilt of its rotational axis, the way it wobbles like a spinning top, and the variable elongation of its orbit around the sun—affect the amount of solar energy received on the surface, and thus potentially influence climate. The variations in each of these parameters are regular: they progress through complete cycles with characteristic—and precisely known—time periods of tens of thousands to hundreds of thousands of years. Milankovitch thought they explained regular variations in climate, especially the comings and goings of ice ages.

In the 1970s, when long sediment cores from the sea floor became available, it was discovered that the chemical properties of ocean sediments accurately record the orbital cycles, suggesting that the similar features in much older sedimentary rocks on land were produced in the same way. And, if that is so, these sedimentary rocks can be used as clocks, because each orbital cycle has an accurately known duration. By measuring just a few radiometric dates spaced at intervals through a sequence of sedimentary rocks, the repeated layer patterns can be used—like tree rings—to count off the time in between, at a resolution approaching tens of thousand of years, well within the paleontologists' requirements.

Better time resolution has also become an important goal for the earliest part of Earth and solar system history, but quite different approaches are necessary for that very distant time in the past. Clair Patterson showed with his uranium-lead data that the Earth—and the meteorites too—date to almost 4.6 billion years ago. But cosmochemists—those who deal with the chemistry and other properties of meteorites and extraterrestrial planets—wanted to know if there might after all be small age differences between meteorites and the Earth, or between different types of meteorites, or even among the different components of individual meteorites. They knew that the solar system didn't reach its present state instantaneously, and that there must have been a formation sequence for the different solid bodies in the solar system, spread over a short but unknown time interval. Patterson's data didn't have the resolving power to discern the small time differences that must have been involved.

Improvements in conventional uranium-lead dating of the sort already described in this book—reductions in contamination, the ability to analyze very small samples—have gone partway toward answering some of the cosmochemists' questions. Thirty years ago, it was thought that the interval over which the sun, planets, and meteorites formed might stretch over 100 million years or more. Now we know that it is an order of magnitude less. With uncertainties in the best cases as low as 1 million years, uranium-lead dates can resolve events that occurred little more than a few million years apart, even if they happened 4.5 billion years ago.

However, discoveries made during the analysis of the calcium-aluminum-rich inclusions (CAIs) described in chapter 5 have shown that much better time resolution is attainable. The CAIs have been dated using the uranium-lead method, but they have also been analyzed for many other isotopes. In these analyses, it was discovered that many of the CAIs, when they first formed, contained radioactive isotopes that are now extinct—they have completely decayed away because of their short half-lives. The only trace left of these extinct isotopes is their daughter products—the stable isotopes into which they were transformed when they decayed. But it is possible to calculate ages from the concentrations of these daughter isotopes, and because the time resolution of radiometric dating methods depends to a large extent on the half-life of the parent—the shorter the half-life, the higher the resolution—cosmochemists realized they might have a way to attain the resolution they were seeking. It would be a different kind of dating exercise, because these particular natural clocks stopped ticking when the now-extinct isotopes completely decayed away and so it would not be possible to calculate absolute ages. But the extinct isotopes offered a way to measure *differences* in absolute ages of early solar system objects precisely and with very high time resolution.

Two especially important now-extinct isotopes that have been used in this way are an isotope of aluminum, aluminum-26, with a half-life of just 0.75 million years, and one of manganese, manganese-53, with a half-life of 3.7 million years. Potentially, these isotopes provide time resolution that is just a fraction of their short half-lives. The daughter products of both these extinct isotopes have been measured in many meteorites and also in individual CAIs, and—although this work is by no means complete—the analyses show that there are short intervals (much less than a million years) between the formation times of different CAIs, even ones from a single piece of the Allende meteorite, such as the one shown in figure 13 (see page 120). The dates also sketch out a kind of cosmic timetable for the formation of different meteorites and meteorite types, spanning a geologically short period of only a few million years. All this fine-scale chronology refers to the very dawn of solar system history, which makes its precise delineation all the more impressive. And while the timing information from the extinct isotopes is in a sense "floating" because it doesn't provide absolute ages, its full potential is realized by using a conventional system such as uranium-lead dating to fix the absolute age of one or more of these events. Each of the relative ages can then be converted to an absolute date, still maintaining the fine-scale time resolution between them.

Thus radioactive isotopes have given us natural clocks with the power to order events with an accurate time scale at both ends of the 4.5-billion-year history of our little corner of the Milky Way galaxy, and at times in between as well. As I said near the beginning of this book, that capability is one of the singular achievements of the earth sciences. In the geological instant that is the last century, a combination of major breakthroughs and smaller improvements in the way we use isotopes to measure time have taken us from having only the vaguest ideas about the magnitude of our planet's history to being able to say with confidence such amazing things as that 5,200 years ago, Oetzi, the Alpine Iceman, died while trekking through the mountains of Europe; that 300 million years ago, what is now southern Africa was covered thick in glacial ice; and that 4,567.2 million years ago, a millimeter-sized grain rich in calcium and aluminum was one of the first, lonely pieces of solid material to condense from the hot gases surrounding our young sun. That is quite an accomplishment.

THE GEOLOGICAL TIME SCALE

The geological time scale shown here is based on the most recent (2004) version from the International Commission on Stratigraphy. I have omitted some of the finer subdivisions of that scale (e.g., epochs of the Mesozoic and Paleozoic, and most subdivisions of the Precambrian) for clarity and because they are not in common use except by specialists. Dates listed are for the beginning of each interval, and are given in years ago (ya), millions of years ago (mya), and billions of years ago (bya). For clarity, many of these dates have been rounded to whole numbers. A few events or features of each subdivision of the time scale are listed.

Eon	Era	Period	Epoch	Begins	Features
Phanerozoic	Cenozoic		Holocene	11,400 ya	N. Hemisphere glaciers retreat
			Pleistocene	1.8 mya	Pleistocene Ice Age begins in N. Hemisphere
		Neogene	Pliocene	5.3 mya	Early hominids; climate cool and dry
			Miocene	23.0 mya	First apes appear
		Paleogene	Oligocene	33.9 mya	Warm climate; mammals diversify rapidly
			Eocene	55.8 mya	First grasses; glaciers form in Antarctic
			Paleocene	65.5 mya	Tropical climate; first large mammals appear
	Mesozoic	Cretaceous		146 mya	Many types of dinosaurs; primitive birds
		Jurassic		200 mya	Giant continent of Pangaea breaks up; conifer and ferns common; dinosaurs
		Triassic		251 mya	First mammals; modern corals appear in oceans
	Paleozoic	Permian		299 mya	Flourishing marine life, 95% extinguished at P-T boundary; continent collision creates Appalacian Mountains
		Carboniferous		359 mya	Primitive trees, first forests; first land vertebrates
		Devonian		416 mya	First (wingless) insects appear; first trees
		Silurian		444 mya	Fish with jaws appear in oceans; first vascular plants on land
		Ordovician		488 mya	Shelled sea life proliferates; first green plants and fungi on land
		Cambrian		542 mya	Great diversification of life in Cambrian explosion; first abundant fossils in sedimentary rocks
Proterozoic		Ediacaran		630 mya	Soft-bodied multicelled life proliferates in oceans
				2.5 bya	Fossils rare; oxygen in atmosphere increases; "Snowball Earth" glaciation near end of eon
Archean				3.8 bya	Oldest probable microfossils
Hadean				4.6 bya	Oldest rocks (Acasta Gneiss) and ancient detrital zircons; life forms likely appeared

PERIODIC TABLE OF THE
CHEMICAL ELEMENTS

When the chemical elements are arranged in order of increasing atomic number (shown to the upper left of each element symbol in the table; this is the number of protons in the nucleus of the element), they exhibit periodic repetition of the same properties. Groups of elements in vertical columns have broadly similar chemical behavior. Two special groups shown at the bottom of the table, the lanthanides (beginning with lanthanum, La) and the actinides (beginning with actinium, Ac), also show similar within-group behavior because of the configuration of electrons around the nuclei of these elements. Members of the actinides with atomic numbers higher than 92 (uranium) are not found naturally on Earth; they are all radioactive and are man-made. Two additional elements in the table, technetium (Tc) and promethium (Pm) are radioactive and not found naturally on Earth.

1 H																	2 He
3 Li	4 Be											5 B	6 C	7 N	8 O	9 F	10 Ne
11 Na	12 Mg											13 Al	14 Si	15 P	16 S	17 Cl	18 Ar
19 K	20 Ca	21 Sc	22 Ti	23 V	24 Cr	25 Mn	26 Fe	27 Co	28 Ni	29 Cu	30 Zn	31 Ga	32 Ge	33 As	34 Se	35 Br	36 Kr
37 Rb	38 Sr	39 Y	40 Zr	41 Nb	42 Mo	43 *Tc*	44 Ru	45 Rh	46 Pd	47 Ag	48 Cd	49 In	50 Sn	51 Sb	52 Te	53 I	54 Xe
55 Cs	56 Ba	57 La	72 Hf	73 Ta	74 W	75 Re	76 Os	77 Ir	78 Pt	79 Au	80 Hg	81 Tl	82 Pb	83 Bi	84 Po	85 At	86 Rn
87 Fr	88 Ra	89 Ac															

57 La	58 Ce	59 Pr	60 Nd	61 *Pm*	62 Sm	63 Eu	64 Gd	65 Tb	66 Dy	67 Ho	68 Er	69 Tm	70 Yb	71 Lu
89 Ac	90 Th	91 Pa	92 U	93 *Np*	94 *Pu*	95 *Am*	96 *Cm*	97 *Bk*	98 *Cf*	99 *Es*	100 *Fm*	101 *Md*	102 *No*	103 *Lw*

Element names corresponding to the chemical symbols are listed below, arranged by increasing atomic number.

1.	**H**	hydrogen	35.	**Br**	bromine
2.	**He**	helium	36.	**Kr**	krypton
3.	**Li**	lithium	37.	**Rb**	rubidium
4.	**Be**	beryllium	38.	**Sr**	strontium
5.	**B**	boron	39.	**Y**	yttrium
6.	**C**	carbon	40.	**Zr**	zirconium
7.	**N**	nitrogen	41.	**Nb**	niobium
8.	**O**	oxygen	42.	**Mo**	molybdenum
9.	**F**	fluorine	43.	**Tc**	technetium
10.	**Ne**	neon	44.	**Ru**	ruthenium
11.	**Na**	sodium	45.	**Rh**	rhodium
12.	**Mg**	magnesium	46.	**Pd**	palladium
13.	**Al**	aluminum	47.	**Ag**	silver
14.	**Si**	silicon	48.	**Cd**	cadmium
15.	**P**	phosphorus	49.	**In**	indium
16.	**S**	sulfur	50.	**Sn**	tin
17.	**Cl**	chlorine	51.	**Sb**	antimony
18.	**Ar**	argon	52.	**Te**	tellurium
19.	**K**	potassium	53.	**I**	iodine
20.	**Ca**	calcium	54.	**Xe**	xenon
21.	**Sc**	scandium	55.	**Cs**	cesium
22.	**Ti**	titanium	56.	**Ba**	barium
23.	**V**	vanadium	57.	**La**	lanthanum
24.	**Cr**	chromium	58.	**Ce**	cerium
25.	**Mn**	manganese	59.	**Pr**	praseodymium
26.	**Fe**	iron	60.	**Nd**	neodymium
27.	**Co**	cobalt	61.	**Pm**	promethium
28.	**Ni**	nickel	62.	**Sm**	samarium
29.	**Cu**	copper	63.	**Eu**	europium
30.	**Zn**	zinc	64.	**Gd**	gadolinium
31.	**Ga**	gallium	65.	**Tb**	terbium
32.	**Ge**	germanium	66.	**Dy**	dysprosium
33.	**As**	arsenic	67.	**Ho**	holmium
34.	**Se**	selenium	68.	**Er**	erbium

69.	**Tm**	thulium		87.	**Fr**	francium
70.	**Yb**	ytterbium		88.	**Ra**	radium
71.	**Lu**	lutetium		89.	**Ac**	actinium
72.	**Hf**	hafnium		90.	**Th**	thorium
73.	**Ta**	tantalum		91.	**Pa**	protactinium
74.	**W**	tungsten		92.	**U**	uranium
75.	**Re**	rhenium		93.	**Np**	neptunium
76.	**Os**	osmium		94.	**Pu**	plutonium
77.	**Ir**	iridium		95.	**Am**	americium
78.	**Pt**	platinum		96.	**Cm**	curium
79.	**Au**	gold		97.	**Bk**	berkelium
80.	**Hg**	mercury		98.	**Cf**	californium
81.	**Tl**	thallium		99.	**Es**	einsteinium
82.	**Pb**	lead		100.	**Fm**	fermium
83.	**Bi**	bismuth		101.	**Md**	mendelevium
84.	**Po**	polonium		102.	**No**	nobelium
85.	**At**	astatine		103.	**Lw**	lawrencium
86.	**Rn**	radon				

APPENDIX C

ADDITIONAL NOTES

I have tried as much as possible to avoid including inherently confusing details and equations in the main text (except for the radiocarbon dating equation on page 79) because they tend to put some readers off. But I realize others may want a bit more information about the principles behind some of the things discussed in this book. For that reason I have briefly elaborated on a few of those topics below.

THE URANIUM DECAY SERIES

All the early explorers of the phenomenon of radioactivity—Becquerel, the Curies, Rutherford—worked with uranium, or with its close neighbor, thorium. What they didn't at first realize, as explained in the main text, is that the other radioactive species associated with uranium and thorium—such as radium and polonium—are actually daughter products of their decay. The radioactive isotopes of uranium and thorium are unusual in the sense that they do not decay directly to a stable daughter product. Instead, they decay through a chain of intermediate isotopes, all radioactive with relatively short half-lives, until a stable isotope of lead is reached. Most of these decays involve emission of an alpha particle from the nucleus of the decaying atom. Alpha particles are actually nuclei of helium atoms, with two neutrons and two protons and therefore an atomic mass of 4. Thus each decay involving alpha particle emission changes the mass of the decaying isotope by 4—e.g., radium-226 decays to

radon-222 by emitting an alpha particle with mass 4. In the decay series starting at uranium-238 (see below), 8 alpha particles are emitted before stable lead-
206 is reached (which you can figure out easily enough: 238 minus 8 times 4, or
32, is 206). Although Rutherford didn't know at first that alpha particles are helium nuclei, he did know that, somehow, helium gas was formed as uranium
decayed, and he used this property in his first attempt to date rocks—he simply measured their helium and uranium contents, and used an estimate of the
helium production rate to calculate an age.

A few of the isotopes in the uranium-238 decay series are shown below.
These are the ones that were of particular interest to early researchers in radioactivity. Similar series begin at thorium-232 and uranium-235; in all three
cases, the end product is a stable isotope of lead. Note the very short half-lives
of the intermediate isotopes compared with uranium.

Uranium-238   (half-life 4.47 billion years)
   ↓

   .

   .   (various intermediate isotopes)

   .
   ↓

Radium-226   (half-life 1,600 years)
   ↓

Radon-222   (half-life 3.8 days)
   ↓

Polonium-218   (half-life 3.1 minutes)
   ↓

   .

   .   (various intermediate isotopes)

   .

Polonium-210   (half-life 138 days)
   ↓

Lead-206   (stable)

THE RADIOACTIVE DECAY EQUATION

Radioactive decay, like many other natural processes, is referred to as a "first-order" process, and it follows simple mathematical rules. Each radioactive isotope decays at a rate that is governed by its decay constant, identified by the Greek letter *lambda* (λ). The decay constant is related to the half-life, as we will see below.

Mathematically, radioactive decay can be descried by an equation that says the number of decays occurring in a particular period of time is proportional to the number of radioactive atoms:

$$-dN/dt \propto N$$

where N is the number of radioactive atoms, t is time, and dN/dt is the instantaneous decay rate of N radioactive atoms. The negative sign is necessary because N decreases with time.

Using calculus, the equation can be integrated to give the form of the radioactive decay equation that is normally used:

$$N = N_o e^{-\lambda t}$$

The decay constant λ appears in the integrated equation, as does the e, representing a constant, the number $2.71828\ldots$, which is the base of natural logarithms. The subscript *zero* (o) refers the value of N on the right-hand side of the equation to its initial condition, when $t = 0$. In words, the equation says that at any time t, the number of radioactive atoms will be equal to the number that were present at $t = 0$ times the expression $e^{-\lambda t}$. This equation describes exponential decay.

The above equation is used "as is" for radiocarbon dating, as shown in chapter 4. The measured quantity, the amount of carbon-14 in a sample, is represented by N. N_0, the carbon-14 content of the sample material when it died, is assumed by agreement to be the same as "modern" carbon for the purposes of calculating a "radiocarbon age," but, in reality, it varied in the past, and the "radiocarbon age" must be adjusted using a calibration curve to obtain the true age of a sample.

By definition, the half-life of any radioactive isotope is the time required for half the initial amount to decay away. The relationship between the decay constant, λ, and the half-life can be calculated easily from the decay equation by setting $N = 0.5\,N_0$. The result is $t_{1/2} = 0.693/\lambda$. Thus either of these constants can easily be calculated from the other.

VARIATIONS ON THE DECAY EQUATION

Only for radiocarbon dating can the decay equation be used in the form shown above. For all of the other methods discussed in this book, a different version is necessary, one that also includes the daughter isotope.

Because each parent atom produces one daughter atom when it decays, the relationship between the two is straightforward. In terms of the decay equation above, the number of daughter atoms *(D)* produced over time *t* would be $N_0 - N$. Thus $D = N_0 - N$, which can be rearranged to $N_0 = D + N$. Substituting for N_0, the decay equation can be rewritten:

$$D = N(e^{\lambda t} - 1)$$

As an example, for the uranium-lead dating method, the equation used to calculate ages (for the uranium-238 to lead-206 decay) would be:

$$^{206}Pb = {}^{238}U(e^{\lambda t} - 1)$$

In this case, the two quantities that must be measured are the amounts of the daughter isotope, lead-206, and the parent isotope, uranium-238. And there is one additional complication. Some minerals may contain small amounts of lead-206 when they form (i.e., at time zero), which, if not taken into account, would invalidate the age calculation because the above equation relates only to the lead-206 produced by radioactive decay. Fortunately, there are ways to get around this difficulty, and it does not present a problem for dating.

Similar equations to the one shown for uranium-238 decay can be written for the other isotope of uranium, uranium-235, and for the potassium-argon and rubidium-strontium dating schemes. The potassium-argon case is slightly more complicated because potassium-40 decays to *both* an isotope of argon (argon-40) and an isotope of calcium (Ca-40). However, the branching ratio is fixed and can be taken into account in the equation.

Among the radiometric methods for age determination, uranium-lead dating has a special place because there are two different isotopes of uranium that decay to two different isotopes of lead. This makes it possible to date samples by measuring only their lead isotopes—no analysis for uranium is required. The rationale can be seen by writing out the equation for uranium-235 decay, which is similar to that for uranium-238 decay shown above:

$$^{207}Pb = {}^{235}U(e^{\lambda t} - 1)$$

If the two equations are divided, one by the other, the result becomes:

$$\frac{^{207}Pb}{^{206}Pb} = \frac{^{235}U(e^{\lambda_5 t} - 1)}{^{238}U(e^{\lambda_8 t} - 1)}$$

To avoid confusion, the decay constants for uranium-235 and uranium-238 are identified by subscripts "5" and "8." The ratio between the two uranium isotopes is fixed (its value is 0.0072). Thus the equation becomes:

$$\frac{^{207}Pb}{^{206}Pb} = \frac{.0072(e^{\lambda_5 t} - 1)}{(e^{\lambda_8 t} - 1)}$$

It is obvious that only the two lead isotopes need to be measured to calculate an age.

GLOSSARY

accretion The term commonly used to describe the process of aggregation of materials to form a planet.

alpha particles (rays) Originally described as "rays," these are actually particles (**nuclei** of helium atoms) consisting of two **neutrons** and two **protons** that are emitted from some isotopes during radioactive decay.

ammonite A commonly fossilized marine mollusk that was abundant during the **Mesozoic** era. It had a coiled and chambered shell similar to the present-day nautilus.

Archean The interval of Precambrian time between 3.8 and 2.5 billion years ago (see appendix A). Derived from the Greek word for "ancient."

atom The basic unit of matter, consisting of a **nucleus** surrounded by **electrons**.

atomic nucleus The central part of an **atom**, where most of its mass resides. It is made up of **protons** and **neutrons**, except for the **isotope** hydrogen-1, in which the nucleus is a single proton.

atomic number The number of **protons** in the **nucleus** of an **atom**; it defines the chemical element.

atomic weight The weight of an **atom** relative to one-twelfth the weight of carbon-12.

beta particles (rays) Originally described as "rays," these are actually particles—they are **electrons** or their positively charged equivalents, positrons.

Cambrian The interval of geological time between 542 and 488 million years ago, characterized by the appearance of animals with shells and other hard

251

parts (see appendix A). The Cambrian gets its name from the classical name for Wales (Cambria), where some of the first detailed studies of rocks of this age were carried out.

cathode ray A stream of **electrons** emitted from the cathode (negatively charged electrode) of a device such as a cathode ray tube.

Cenozoic From the Greek words for "new" and "animal" or "life," the interval of geological time between 65.5 million years ago and today, characterized by the rise in importance of the mammals (see appendix A).

cosmic rays High-energy particles (the **nuclei** of **atoms**) that reach the Earth from outside the solar system; when they collide with atoms of the Earth's atmosphere, they often produce additional particles (such as **neutrons** and **protons**) that are referred to as secondary cosmic rays.

cyanobacteria A phylum that includes all the photosynthesizing bacteria, often referred to as blue-green algae.

DNA The common name for deoxyribonucleic acid, the molecules of which contain the genetic information in nearly all organisms, with the exception of some viruses.

electrometer A sensitive instrument for measuring very small electric currents or voltages.

electron A small particle that carries a negative electric charge. Electrons surround the **nucleus** in **atoms** and balance the positive charge of the **protons.** They are the primary carrier of electricity in conductors.

fluorescence The phenomenon of light emission from **atoms** when they are excited by short-wavelength radiation such as ultraviolet or **X-rays.**

gamma ray A form of energetic electromagnetic radiation produced when **atomic nuclei** shift from one energy level to another.

gneiss A variety of **metamorphic rock** characterized by minerals that tend to be flattened out in a single direction, giving the rock a banded appearance.

granite A course-grained **igneous rock** that cooled and crystallized at depth in the Earth's crust. It is composed mostly of the minerals quartz, feldspar, and mica.

graphite A crystalline form of carbon that is stable at low temperatures and pressures.

Hadean From the Greek and usually referring to the underworld or hell, the Hadean comprises the first interval of the Earth's history from its formation to the beginning of the **Archean,** 3.8 billion years ago (see appendix A). It is often depicted as a time when the Earth was very hot; hence the name.

half-life The time required for half of a quantity of radioactive material to decay away.

hominid A member of the biological "great apes" family, which includes gorillas, chimpanzees, orangutans, and humans. Use of the term is quite varied; originally it referred only to *Homo sapiens* and their immediate fossil precursors, and it is still used in this way by some.

ice age A period during which average temperatures on Earth were low and substantial portions of the continents were covered with thick glaciers. There have been a number of ice ages throughout the Earth's long history, but in this book the term is used primarily with reference to the most recent such event, which started about 2 million years ago in the Northern Hemisphere and is often referred to as the **Pleistocene** Ice Age (see appendix A). The Pleistocene Ice Age has been characterized by long cold periods interrupted by shorter, warmer "interglacials." Today's climate is generally agreed to be an interglacial within the continuing Pleistocene Ice Age.

igneous rock Rock that is formed by the cooling and crystallization of molten magma.

ion An **atom** with an electric charge, either positive or negative, produced by a deficit or excess of **electrons**.

isotope Any of a set of **atoms**, the **nuclei** of which have the same number of **protons** but different numbers of **neutrons**. They are therefore the same chemical element but have slightly different masses.

magma Melted rock, formed in the Earth's interior.

Mesozoic From the Greek words for "middle" and "animal" or "life." The Mesozoic lasted from 251 to 65.5 million years ago (see appendix A). Both its beginning and its end are marked by massive extinctions of life on Earth.

metamorphic rock A rock formed when preexisting rocks are subjected to new environmental conditions, usually higher temperatures and/or pressures. Typically, metamorphism involves changes in the mineral makeup and structure of the original rocks.

microfossil A fossil so small that it can only be examined with a microscope.

neutron A neutral particle that is an integral component of all **atomic nuclei** (except for hydrogen nuclei, which contain only a proton). In the free state (i.e., when not inside a nucleus), neutrons are unstable, with a **half-life** of 15.2 minutes.

nuclear fission A type of radioactive decay in which a heavy (large) **atomic nucleus** breaks up into two major fragments and several smaller ones.

nuclear fusion The combination of lighter **atomic nuclei** to make heavier ones, with the release of large amounts of energy. The fusion of hydrogen nuclei to produce helium is the reaction that powers our sun.

nucleus See **atomic nucleus.**

oolite A type of **sedimentary rock** composed of small (typically 1 millimeter or less in diameter) spheroidal particles of calcium carbonate. It forms in warm, shallow-water environments.

outcrop An exposed rock unit or stratum.

Paleozoic From the Greek words for "ancient" and "animal" or "life." The Paleozoic began with the beginning of the **Cambrian** period 542 million years ago, and ended in the great mass extinction that occurred 251 million years ago (see appendix A).

periodic table The chemical elements arranged in periods (rows) and groups (columns) of increasing **atomic number,** illustrating common chemical behavior among groups (see appendix B).

Permian The last period of the **Paleozoic** era, spanning the interval between 299 and 251 million years ago (see appendix A). It ended with the greatest mass extinction the Earth has known. The Permian is named after exposures of **sedimentary rocks** of this age near the city of Perm, in the Ural Mountains area of Russia.

Phanerozoic From the Greek words for "visible" or "evident," and "life." The Phanerozoic eon of geological time encompasses the interval from the beginning of the **Cambrian** period, 542 million years ago, until the present (see appendix A).

photosynthesis The process by which plants make carbohydrates from water and carbon dioxide, using the energy supplied by natural light.

pitchblende An ore of uranium.

plankton A general term for the organisms that are free floating (not swimming) in bodies of both fresh and salt water.

plate tectonics A theory that explains many large-scale phenomena in geology, in which plates (many tens to more than a hundred miles thick) make up the outermost part of the Earth. In response to very slow convection in the hot interior of the Earth, the plates move about relative to one another at speeds that are typically an inch to a few inches per year. Many geological features and phenomena, from volcanoes and earthquakes to mountain ranges and mineral deposits, are linked to the interactions between plates.

Pleistocene The interval between 1.8 million years ago and 11.4 thousand years ago (see appendix A). Traditionally, the Pleistocene was thought to encompass the present **ice age**, from the time of its beginning until the waning of the Northern Hemisphere glaciers about 11 thousand years ago. However, the beginning of the Pleistocene Ice Age has now been pushed back somewhat beyond 1.8 million years ago. The name originated with Charles Lyell, who divided up recent geological time on the basis of the percentage of fossils that were recognizable as living species (as outlined in chapter 6). Pleistocene, from the Greek, means "fullest"; in Lyell's classification, it contained the greatest fraction of fossils that could be matched to living animals.

Pliocene The interval between 5.3 and 1.8 million years ago; it immediately precedes the **Pleistocene** (see appendix A). The name originated with Charles Lyell (see **Pleistocene**) and is from the Greek word meaning "full."

Proterozoic The Precambrian interval between 2.5 billion years ago and the beginning of the **Cambrian** period 542 million years ago (see appendix A). From the Greek words for "first," "former," or "first in time," and "animal"; the earliest complex fossil organisms appear toward the end of the Proterozoic.

proton An integral particle in all **atomic nuclei**. A proton has a positive electric charge equivalent to the negative charge on the **electron**.

radioactivity The phenomenon through which unstable **nuclei** are transformed by emission of particles and/or radiation.

radiocarbon The radioactive **isotope** of carbon, carbon-14.

Richter scale A scale used to describe the intensity of earthquakes. Great earthquakes are those that register 8.0 or above on the Richter scale. Each increase of one unit on the scale corresponds to an increase of approximately thirty-two times in the energy released by the earthquake.

RNA The common name for ribonucleic acid, a molecule that is involved in transferring genetic information from **DNA** and in synthesizing proteins.

sedimentary rock A rock formed by the consolidation of loose sedimentary particles, usually due to burial and cementation, or by direct chemical precipitation from water (for example, "rock salt"). Sedimentary rocks form both on land and underwater.

stratigraphy The study of **sedimentary rock** sequences, especially their distinguishing characteristics such as mineral makeup, chemical composition, and fossil content.

subduction zone The region where an oceanic plate collides with and plunges under the edge of a continental plate (see **plate tectonics**).

till The loose, unsorted debris of a glacial deposit.

trilobite A marine arthropod common during the **Paleozoic** era. It had a segmented, three-part body.

tsunamis Great ocean waves generated when earthquakes, or sometimes landslides, displace large volumes of water. Tsunamis have long wavelengths (up to 150 miles) and travel at high speeds. They can reach great heights (more than 50 feet) when they interact with shoaling coastlines or are funneled into bays.

unconformity A break in a sequence of **sedimentary rocks**, often indicating a long time period with no deposition and/or active erosion. Observable as the contact between two rock units of quite different ages and often different types or orientations.

X-ray Short-wavelength electromagnetic radiation.

zircon A mineral composed of the elements zirconium, silicon, and oxygen (its chemical formula is $ZrSiO_4$). It is part of the silicate mineral family, and is extremely resistant to alteration under a wide range of chemical and physical conditions.

RESOURCES AND
FURTHER READING

Below I have listed, by chapter, some of the books and articles I consulted in preparing this book, as well as others that might be of interest to readers who want to explore the subject of radiometric dating and geological time more deeply. Quite a few works are technical articles from scientific journals; most of these can be found in any university library or, in some cases, as electronic versions on the Internet. Books by historical figures such as James Hutton are long out of print and difficult to find unless you have access to a research library, but, in most cases, there are plenty of secondary sources, some of them listed here. A great place to start is the Internet, especially sites like Wikipedia, the free on-line encyclopedia. Most of the topics touched on in this book can be found there, with references and links if you want to delve further. Generally the Wikipedia articles are of high quality, but it is always worthwhile to check details using several sources.

CHAPTER 1

Dalrymple, G. Brent.
    1991.   *The age of the Earth.* Stanford, CA: Stanford University Press. A good general treatment of radioactivity and many of the topics covered in this book. Some of the discussions (e.g., the Earth's oldest rocks) are now a bit dated.
Faure, Gunter, and Teresa M. Mensing.
    2005.   *Isotopes: Principles and applications.* 3rd ed. Hoboken, N.J.: Wiley.

A standard textbook that covers all aspects of radiometric dating, as well as many other uses of isotopes in the earth sciences.

Gould, Stephen Jay.

1987.   *Time's arrow, time's cycle.* Cambridge, MA: Harvard University Press. A scholarly treatment of ideas about geological time.

Hutton, James.

1795.   *Theory of the Earth, with proofs and illustrations.* Edinburgh: William Creech. Hutton's multivolume tome outlining his ideas.

Price, Derek de Solla.

1974.   Gears from the Greeks. *Transactions of the American Philosophical Society* 64: 1–70. An essay about the Antikythera mechanism. Price's work has recently been revised and extended based on new results from sophisticated imaging analyses of the device, as described in several articles in *Nature* 444 (30 November 2006).

Repcheck, Jack.

2003.   *The man who found time.* Cambridge, MA: Perseus Publishing. A great little book about James Hutton and his time (no pun intended).

CHAPTER 2

Goldsmith, Barbara.

2005.   *Obsessive genius: The inner world of Marie Curie.* New York: Norton. An interesting treatment of Marie Curie's life, focusing on the inner qualities that drove her to succeed.

Wilson, David.

1983.   *Rutherford: Simple genius.* London: Hodder and Stoughton. A comprehensive and admiring portrait of Rutherford's life and accomplishments.

CHAPTER 3

Arnold, Jim, and Ernie Anderson.

1996.   Interview by R. E. Taylor. This interview deals with the development of radiocarbon dating in Libby's laboratory and is the source for some of the personal anecdotes described in this chapter. It can be accessed through the special collections section of the University of California at San Diego library under call number SPL-1305A.

Libby; Willard F.

1955.   *Radiocarbon dating.* 2nd ed. Chicago: University of Chicago Press. A small gem of a book, first published in 1952. Libby clearly and succinctly covers the principles and possible applications of the radiocarbon dating method.

CHAPTER 4

Arnold, J. R., and W. F. Libby.

1949.   Age determinations by radiocarbon content: Checks with samples of known age. *Science* 110: 678–80. The famous paper reporting the radiocarbon dates for samples of known age.

1951.   Radiocarbon dates. *Science* 113 : 111–20. The first in a series of five papers reporting yearly updates of the ages measured in Libby's lab. The other four are authored by Libby alone (see below).

Libby, W. F.

1951.   Radiocarbon dates, II. *Science* 114: 291–96.

1952.   Chicago radiocarbon dates, III. *Science* 116: 673–81.

1954a.   Chicago radiocarbon dates, IV. *Science* 119: 135–40.

1954b.   Chicago radiocarbon dates, V. *Science* 120: 733–42.

CHAPTER 5

Bowring, S., I. S. Williams, and W. Compston.

1989.   3.96 Ga gneisses from the Slave Province, Northwest Territories, Canada. *Geology* 17: 971–75. Reports the first dates that revealed the great age of the Acasta Gneiss, the world's oldest rock complex.

Froude, D. O., T. R. Ireland, P. D. Kinny, I. S. Williams, W. Compston, I. R. Williams, and J. S. Myers.

1983.   Ion microprobe identification of 4,100–4,200 myr-old terrestrial zircons. *Nature* 304: 616–18. The first hint that zircon crystals from the first half billion years of the Earth's history had survived.

Lewis, Cherry.

2000.   *The dating game: One man's search for the age of the Earth.* New York: Cambridge University Press. The life of Arthur Holmes.

Patterson, Clair.

1956.   Age of meteorites and the Earth. *Geochimica et Cosmochimica Acta* 10: 230–37. The famous paper in which Patterson established a close connection between the Earth and meteorites, and determined the age of the Earth.

1995.   Interview by Shirley K. Cohen. Oral History Project, California Institute of Technology Archives, Pasadena, California. The text of the interview is available at http://resolver.caltech .edu/CaltechOH:OH_Patterson_C.

CHAPTER 6

Berry, William B. N.

1987.   *Growth of a prehistoric time scale: Based on organic evolution.* Palo Alto, CA: Blackwell Scientific Publications. Berry traces the origins of the modern geological time scale, with detailed discussions of most major subdivisions.

Bowring, S., D. H. Erwin, Y. G. Jin, M. W. Martin, K. Davidek, and W. Wang.

1998   U/Pb zircon chronology and tempo of the end-Permian mass extinction. *Science* 280: 1039–45. A detailed study of the timing of the P-T boundary, mostly based on work on volcanic ash layers from Meishan, China.

Erwin, Douglas.

2006.   *Extinction: How life on Earth nearly ended 250 million years ago.* Princeton, NJ: Princeton University Press. All the relevant details about the P-T extinctions, including dating.

Geological time scale.

     Go to the Wikipedia main page (http://en.wikipedia.org/wiki/ Main_Page) and search for "geological time scale" to view an up-to-date version with more information than you probably want.

Schneer, Cecil.

     William "Strata" Smith on the Web. A website that includes explanatory text about Smith as well as reproductions of his map and his fossil illustrations. The website is the work of Professor Cecil Schneer of the University of New Hampshire and can be accessed at www.unh.edu/esci/wmsmith.html.

Winchester, Simon.
    2001.   *The map that changed the world: William Smith and the birth of modern geology.* New York: Harper Collins. An interesting book about William Smith's life and times, and the making of the first geological map.

CHAPTER 7

Alvarez, L. W., W. Alvarez, F. Asaro, and H. V. Michel.
    1980.   Extraterrestrial cause for the Cretaceous-Tertiary extinction. *Science* 208: 1095–1108. The original paper announcing the evidence for an impact.

Alvarez, Walter.
    1997.   *T. rex and the crater of doom.* Princeton, NJ: Princeton University Press. A book about the K-T impact and extinctions, written by one of the scientists who discovered the "iridium anomaly" that implicated a large impact in the mass extinctions.

Gould, Stephen J.
    1994.   The evolution of life on Earth. *Scientific American,* October, 85–91. An overview by one of the most prominent thinkers in the field. Although there have been many new developments since this was written, it is still a very good guide.

Manning, Craig, Stephen J. Mojzsis, and T. Mark Harrison.
    2006.   Geology, age, and origin of the supracrustal rocks at Akilia, West Greenland. *American Journal of Science* 306: 303–66. The paper giving a minimum age and evidence for sedimentary origin of the Greenland rocks that contain graphite of possible biologic origin.

Schopf, J. William.
    1999.   *Cradle of life: The discovery of the Earth's earliest fossils.* Princeton, NJ: Princeton University Press. Although there is considerable controversy about whether the "earliest fossils" identified by Schopf are really biological in origin, this is a good exposition of Schopf's arguments (and an interesting story of the discoveries).

CHAPTER 8

Atwater, Brian F., Musumi-Rokkaku Satoko, Satake Kenji, et al.

2005. *The orphan tsunami of 1700: Japanese clues to a parent earthquake in North America.* U.S. Geological Survey paper no. 1707. Reston, VA: U.S. Geological Survey; Seattle: University of Washington Press. A beautifully illustrated account of the story behind finding a North American earthquake source for the tsunami that struck Japan in 1700. A PDF version of this book is available for download from the U.S. Geological Survey at http://pubs.usgs.gov/pp/pp1707.

Damon, P. E., D. J. Donahue, B. H. Gore, et al.

1989. Radiocarbon dating of the Shroud of Turin. *Nature* 337: 611–15. The paper reporting the radiocarbon dates from three accelerator mass spectrometry laboratories for the Shroud of Turin.

Fairbanks, Richard G.

Fairbanks 0107 calibration curve. A calibration curve proposed by Richard Fairbanks of the Lamont-Doherty Earth Observatory of Columbia University, New York. Fairbanks's curve extends back to 50,000 years before the present and is used by many researchers. It is available on Fairbanks's "Current Research" website, www.radiocarbon.ldeo.columbia.edu/research/radcarbcal.htm.

Friedrich, Walter L., Bernd Kromer, Michael Friedrich, et al.

2006. Santorini eruption radiocarbon dated to 1627–1600 B.C. *Science* 312: 548. Reports the dating of an olive branch buried in volcanic ash from the Santorini eruption.

Kelsey, Harvey M., Alan R. Nelson, Eileen Hemphill-Haley, and Robert C. Witter.

2005. Tsunami history of an Oregon coastal lake reveals a 4600 yr record of great earthquakes on the Cascadia subduction zone. *Bulletin of the Geological Society of America* 117: 1009–32. Reports results from a study of sediment cores from Bradley Lake.

Reimer, P. J., M. G. L. Baillie, E. Bard, et al.

2004. IntCal04 terrestrial radiocarbon age calibration, 0–26 cal kyrs B.P. *Radiocarbon* 46: 1029–58. The "official" radiocarbon calibration curve for this time period. The paper gives details of the data that went into the final version.

CHAPTER 9

Elmore, David E., and Fred M. Phillips.

1987.   Accelerator mass spectrometry for measurement of long-lived radioisotopes. *Science* 236: 543–50. Although published more than twenty years ago, this is still a good summary of the principles of the method and some of its applications. A more up-to-date and technically comprehensive (and very expensive) treatment can be found in C. Tuniz, J. R. Bird, D. Fink, and G. F. Herzog, *Accelerator mass spectrometry: Ultrasensitive analysis for global science* (Boca Raton, FL: CRC Press, 1998).

Erwin, Douglas H.

2006.   Dates and rates: Temporal resolution in the deep time stratigraphic record. *Annual Review of Earth and Planetary Sciences* 34, 569–90. A recent summary of some of the issues surrounding high-resolution dating discussed in this chapter.

Lugmair, G. W., and A. Shokolyukov.

2001.   Early solar system events and timescales. *Meteoritics and Planetary Sciences* 36: 1017–26. A very good summary of how short-lived extinct isotopes have provided a high-resolution chronology of events in the early solar system.

INDEX

Text: 11/15 Granjon
Display: Granjon
Compositor: Binghamton Valley Composition
Printer and binder: Maple-Vail Manufacturing Group